Lecture Notes in Earth Sciences

88

Editors:
S. Bhattacharji, Brooklyn
G. M. Friedman, Brooklyn and Troy
H. J. Neugebauer, Bonn
A. Seilacher, Tuebingen and Yale

W0091210

Springer
Berlin
Heidelberg
New York
Barcelona
Hong Kong
London
Milan
Paris
Singapore
Tokyo

Saulo Rodrigues-Filho   German Müller

A Holocene Sedimentary Record from Lake Silvana, SE Brazil

Evidence for Paleoclimatic Changes
from Mineral, Trace-Metal
and Pollen Data

With 29 Figures and 11 Tables

Springer

Authors

Dr. Saulo Rodrigues-Filho
Geochemist
CETEM/CNPq, Centro de Technologia Mineral
Rua 4, Quadra D. Cidade Universitaria, 21941.590 Rio de Janeiro, Brazil

Prof. Dr. Dr. h. c. mult. German Müller
Ruprecht-Karls-Universität Heidelberg, Institut für Umweltgeochemie
Im Neuenheimer Feld 236, D-69120 Heidelberg, Germany

Cataloging-in-Publication data applied for

Die Deutsche Bibliothek - CIP-Einheitsaufnahme

Rodrigues-Filho, Saulo:
A holocene sedimentary record from Lake Silvana, SE Brazil :
evidence for paleoclimatic changes from mineral, trace metal and
pollen data / Saulo Rodrigues-Filho ; German Müller. - Berlin ;
Heidelberg ; New York ; Barcelona ; Hong Kong ; London ; Milan ;
Paris ; Singapore ; Tokyo : Springer, 1999
  (Lecture notes in earth sciences ; 88)
  ISBN 3-540-66205-7

"For all Lecture Notes in Earth Sciences published till now please see final pages of
the book"

ISSN 0930-0317
ISBN 3-540-66205-7 Springer-Verlag Berlin Heidelberg New York

Typesetting: Camera ready by author
SPIN: 10736938   32/3142-543210 - Printed on acid-free paper

To Andréa Vilhena and Júlia and Pedro Vilhena Rodrigues

Preface

In 1950 Martin Schwarzbach from Cologne University published a remarkable book „Das Klima der Vorzeit„, followed by the second (1961) and third (1974) edition and an English version „Climates of the Past„ (1963), tracing the historical evolution of the earth's climate since Cambrian time.

The catalytic effect of his book was enormous and encouraged worldwide research with the development and application of new methods, to mention only isotopic methods to reconstruct paleotemperatures.

During the past two decades the climate question has become a new *ecological* dimension: Will the „greenhouse effect„, related to the combustion of fossil fuels, influence our *present* climate? Will it remain as it is, will it become colder, or will it become warmer? And: how rapid will a change occur (if it occurs at all!)?

Studies on paleoclimatic development to-day concentrate on Pleistocene sediments and on ice cores deposited during and between glacial transgressions and on post-glacial sediments laid down during the Holocene.

The present study is an example of a sediment series in a lake formed some 10.000 years ago in a now tropical climate. Already by vision four different sediment types can be recognized which represent four paleoenvironmental zones. Mineralogical-sedimentological, geochemical and palynological investigations permit the establishment of a general climatic change from grassland to savanna type vegetation and a period of stronger rainfall leading to the present-day semideciduous forest.

The study is also an example for a multidisciplinary approach which permits the connection of sediment data with the weathering and erosion history in the catchment.

Acknowledgements

We would like to thank the CAPES (Brazilian Coordination for Post-Graduation) for the financial support with a fellowship, the Institute of Environmental Geochemistry at the University of Heidelberg for analytical facilities and financial support, and the CETEM/CNPq (Center for Mineral Technology) for logistical support during the field work in Brazil.

The first author is greatly indebted to Prof. Dr. Roberto C. Villas Bôas for his support in drawing the outlines of this project, as well as to Dr. Fernando A. Freitas Lins and Adão Benvindo da Luz for the institutional support from CETEM/CNPq.

We are grateful to Dr. Hermann Behling, who kindly carried out the pollen analyses in lake sediments. Without the participation of Prof. Dr. Georg Irion in the sampling campaign, the coring operation would not have been possible.

Suggestions by Dr. Jörg Matschullat helped a great deal during the preparation of this manuscript. We are also grateful to Anne Marie De Grosbois for revising the text.

The analytical support provided by Silvia and Stephan Rheinberger was a valuable help. We are pleased to acknowledge the friendly collaboration by Manfred Gastner in the mineralogical analyses.

Dr. Hans-Peter Meyer helped in operating the X-ray fluorescence spectrometer. Dagmar Eggergluss and Ralf Ottenstein helped with chemical analyses and data processing.

The first author would like to express his special thanks to his wife, Andréa M. Gouthier de Vilhena and their children, Júlia and Pedro Vilhena Rodrigues, as well as to his parents, Saulo Rodrigues Pereira and Célia F. Rios Pereira, for the long-term and valuable support.

Contents

1 Abstract

Geochemical, mineralogical and palynological data from a 12.7-m-long sediment core demonstrate the relevance of an interdisciplinary approach to paleoenvironmental studies. The history of the Holocene sedimentation in Lake Silvana was reconstructed by tracing back the source of detrital sediments based on contrasting geochemical and mineralogical patterns within the pedologic mantle of the catchment. Pollen record has led to the reconstruction of the paleovegetation, while comparison of the geochemical and mineralogical composition of lake sediments with that from the weathering mantle provided evidence of changes in lake level and slope-erosion activity since the early Holocene. Such evidence yields some insights into the Holocene landscape evolution and the origin of the lake itself. The lowermost core section reached the Holocene-Pleistocene boundary, according to radiocarbon age determinations with accelerator mass spectrometry (AMS). Manganese and Fe concentrations are shown to be good tracers for the formation of the authigenic component in the lake, while the allogenic component was well traced by Al, Ti and Pb concentrations.

Four major paleoenvironmental zones indicative of distinct phases of the hydrological cycle in Lake Silvana are consistent with vegetation changes indicated by pollen data since the early Holocene. In zone I, yellowish silty-clayey sediments present trace-metal concentrations and a mineralogical composition that resemble those from the lower weathering profile, indicating an environment characterized by low water level and oxic conditions. In zone II, the occurrence of well-sorted clayey sediments composed of kaolinite, gibbsite, goethite and siderite, and with slightly elevated concentrations of Mn and organic C, suggests anoxic conditions and the formation of a shallow lake. Sediment geochemistry and pollen data point to a very early stage of the climate-driven geomorphic process that culminated in the formation of Lake Silvana.

The rapid sedimentation of a 7.8-m-thick sequence of slope-wash sediments, named zone III, appears to be related to the formation of Lake Silvana and the whole lake system of the middle Rio Doce Valley, southeastern Brazil. Slope erosion has been favored by a pollen-indicated sparse vegetation (tropical savanna). Events of slope instability suggest sharply increased precipitation and correspond to two pollen-indicated climatic transitions from dry to moister conditions that culminated 8500 years ago. This age resembles that of forest expansion registered in southern and southeastern Brazil and in southeastern Africa, as well as the early to mid-Holocene climatic event observed in records from ice cores. The records from Greenland point to an inverse fluctuation relative to that indicated for the study area, that is a notable increase in aridity, probably reflecting climatic changes in the Northern Hemisphere, as the bigger size of the northern continents suggests. Increasing cold and dry conditions from 9000 to 6000 years B.P. in northern low latitudes have been explained elsewhere with the orbitally induced decrease in summer insolation (July to August), which would have the opposite effect in the southern tropics.

For the phase preceding the natural damming that likely gave rise to Lake Silvana, represented by zone I, sediment composition indicates a source from lower horizons of the weathering mantle. After this phase, materials derived from the uppermost pedologic horizon started to accumulate in the lake basin, thus forming the zone II. For the phase of intense slope-erosion activity, represented by zone III, detrital sediments are likely derived from the mixing of horizons regardless of their topographic position. In turn, sediments deposited during phases of less intense erosion and lacustrine sedimentation show no evidence of source, as observed for zone IV, probably as a result of their major authigenic nature.

Since 8500 years B.P., environmental changes were marked by the expansion of forest and increasing proportions of the authigenic component in sediments, which characterize the whole zone IV. The vertical distribution of siderite-rich layers, arranged at intervals of ~1 meter, points to the cyclic recurrence of episodes characterized by increasing seasonality with abnormally strong summer rain. These episodes possibly recurred at intervals of 2000-2500 years, assuming that sedimentation rates did not undergo significant variations within zone IV. If this is true, there is a noticeable correspondence in time with cyclic climatic events recorded from Greenland ice cores, which indicate accentuated cold and dry climates in the Northern Hemisphere. Once this hypothesis is confirmed, one may realize the suitability of using siderite as a potential paleoenvironmental indicator for tropical lakes, which is favored by its very limited thermodynamic field of stability.

A high average concentration of Hg (0.20 µg/g) for the uppermost unit of the weathering profiles (colluvial unit) represents a 10-fold increase relative to the average concentration of the saprolite. However, a weathering-driven Hg enrichment is unlikely, since the ionic potential of Hg and other observations from tropical regions point to its high mobility under oxidizing conditions. The indicated age for tropical lateritic soils, ~ 5 Ma., suggests that a long-term deposition of atmospheric Hg derived from natural emissions is likely to explain this Hg accumulation rather than a relative enrichment from the parent rock. The Hg record from the sediment core shows elevated concentrations as old as 9000 years B.P., excluding the hypothesis of Hg derived from anthropogenic emissions.

The main variables controlling the metal distribution, possibly except for Hg, within the weathering profile EG2 are the composition of the parent rock and the degree of weathering. The increase in gibbsite contents at the surface is a typical picture of highly leached soils, while high concentrations of Ti, Cr, Pb, Ni, Cu and Zn in the saprolite horizons seem to be mainly influenced by the mineralogical composition of the parent rock.

2 Introduction

The number of paleoclimatological studies in tropical regions is still limited and just a few of them use the correlation of geological and biological data for reconstructing paleoclimatic changes (e.g., Williamson et al., 1998; Salgado-Labouriau et al., 1997; Johnson et al., 1996). Even for temperate lakes, the use of multiple paleoenvironmental proxies is somewhat restricted (e.g., Matschullat et al., 1998; Yu and Eicher, 1998; Valero-Garcés et al., 1997; Rosenbaum et al., 1996). Investigations on paleoclimates are commonly based either on biological records, such as pollen, macrofossils and tree rings or on geochemical/geophysical records, which include principally oxygen isotope and magnetic susceptibility data. The mineral matrix for pollen and plant macrofossils in lake sediments is commonly not taken into consideration by most paleoclimate studies.

The aim of this study is to investigate the relationship between a 12.7-m-long sediment core and weathering profiles of the Lake Silvana's catchment with regard to their geochemical and mineralogical composition. This relationship is believed to yield information to which extent variations in sediment composition within the sediment core reflect changes in lake level and erosion activity on catchment soils, which in turn are a function of the climate-driven rainfall regime and vegetation type. To test this hypothesis, a palynological record from the sediment core is compared with the sediment/soil-inferred climate evolution.

The subordinate use of geochemical and mineralogical data as paleoenvironmental indicators is probably derived from difficulties in determining the source of the mineral matter found in lake sediments, as well as from early-diagenetic chemical changes that take place in sediments after deposition (Engstron and Wright Jr., 1984). Based on observations from temperate regions, these authors pointed out that definitive evidence for soil erosion may be difficult to obtain from sediment chemistry because changes in soil mineralogy are too subtle to detect. However, this seems not to be the case in tropical regions. Here, chemical weathering causes marked mineral transformations and the relative accumulation of less mobile metals in surface horizons under neutral and oxidizing conditions, such as Al, Ti, Fe, Mn, Cr, Ni, Mo, Nb, Be, Zr, V, U and Th. This provides a potential tool for reconstructing the history of erosion activity in the catchment from the chemical stratigraphy of lake sediments. Figure 2.1 illustrates the relationship between ionic potential, expressed by the ionic charge/ionic-radius ratio, and the mobility of chemical elements, where immobile major and trace elements tend to form hydroxides during weathering processes (Goldschmidt, 1937 in Valeton, 1972).

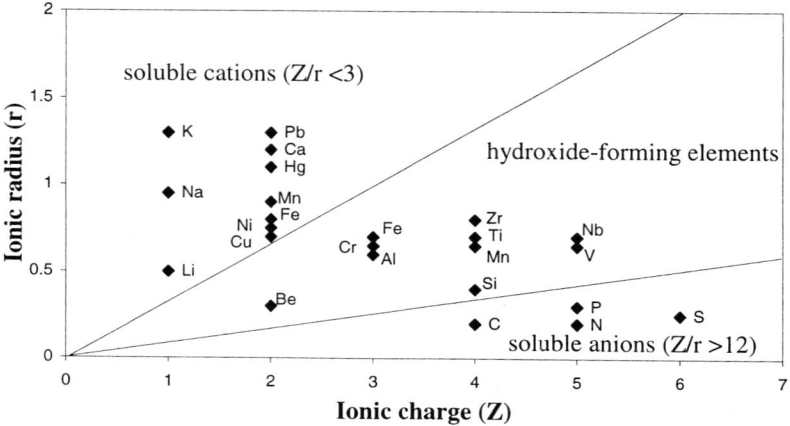

Figure 2.1 Element mobility as a function of ionic potential (Goldschmidt, 1937, modified after Valeton, 1972).

Another difficulty in establishing the signature of a particular soil horizon from the chemical and mineralogical composition of sediments has been pointed out by Chesworth (1972). According to this author, erosion and transport would inevitably destroy many, if not all, pedogenic features either through sorting, in the case of slow sedimentary processes, or through a rapid and indiscriminate sedimentation that tends to homogenize pedological contrasts. Although this argument appears reasonable, another interpretation of the effects of different erosional patterns can be formulated, according to which soil-derived imprints could be recognized in sediments: if sedimentation is slow, detrital material tends to stem from superfical horizons as a result of less intense erosion activity, while during more intense erosion, such as slope wash or debris flow, sediment composition will likely be influenced by soils from deeper horizons.

The contrasting accumulation of quartz and secondary minerals, such as kaolinite, gibbsite, hematite and goethite, within profiles of the weathering mantle in tropical regions is well documented (Lucas et al., 1993; Nahon, 1991; Kopp, 1986; Irion, 1984; Curi and Franzmeier, 1984). Goethite and gibbsite generally occur in the uppermost horizon of profiles as a result of hydration of hematite and leaching of silica from kaolinite, respectively. Moreover, Kopp (1986), Irion (1984) and Kronberg et al. (1979) have observed marked contrasts in the vertical distribution of several trace elements in different weathering profiles throughout tropical Brazilian regions. The most common elements showing a relative enrichment upwards were Al, Fe, Ti, Cr, Ni and V. Therefore, contrasting vertical distribution of less mobile metals and secondary minerals within the weathering mantle may favor the reconstruction of the history of erosion activity, since the composition of the sediment core presents relevant variations.

The purpose of identifying the source of detrital sediments in Lake Silvana also requires the recognition of the authigenic sedimentary component. In this study, the authigenic component is considered those materials formed *in situ* either from the dissolved and organically bound phase or from clastics that enter the lake and undergo early-diagenetic transformations within the sediments. The abundance of minerals and metallic compounds formed within the sediment/water column of Lake Silvana has been assessed from fluctuations in the contents of organic matter, and Fe and Mn behavior relative to the record of conservative elements, such as Al and Ti.

2.1 Study Area

The Rio Doce Valley is situated in the eastern part of the State of Minas Gerais, southeastern Brazil, in one of the world´s most important mining regions for iron. Both Rio Doce and Rio Piracicaba form the so-called *Steel Valley*, where three steel-producing companies have been in operation for the past 40–50 years. Gold mining activities exist in the region since the late 18th century. Heavy-metal pollution derived from the steel industries in waterways of the *Steel Valley* has been reported for Cr, Cd and Pb (Jordão et al., 1996). No geochemical or environmental study dealing with the historical variation of heavy-metal concentrations in sediments is known for this area.

Rio Doce is a 750-km-long river draining a catchment area of 83,000 km^2. Its middle course comprises an area of more than 1500 km^2, where a lake system with almost 40 closed-basin lakes exists (Pflug, 1969; Tintelnot, 1995). The study area includes Lake Silvana (19° 31' S and 42° 25' W) and its catchment, situated on the right bank of Rio Doce, approximately 10 km downstream from the town of Ipatinga (Fig. 2.2).

Lake Silvana is a relatively shallow lake (maximum depth 10 m) with a surface area of approximately 4.6 km^2. The natural vegetation around the lake is formed by semideciduous forest. The climate is warm, semihumid with a dry season of 4–5 months, and a mean annual temperature around 20°C. The annual precipitation is concentrated in the summer – December to March – reaching an average of 1250 mm (Nimer, 1989).

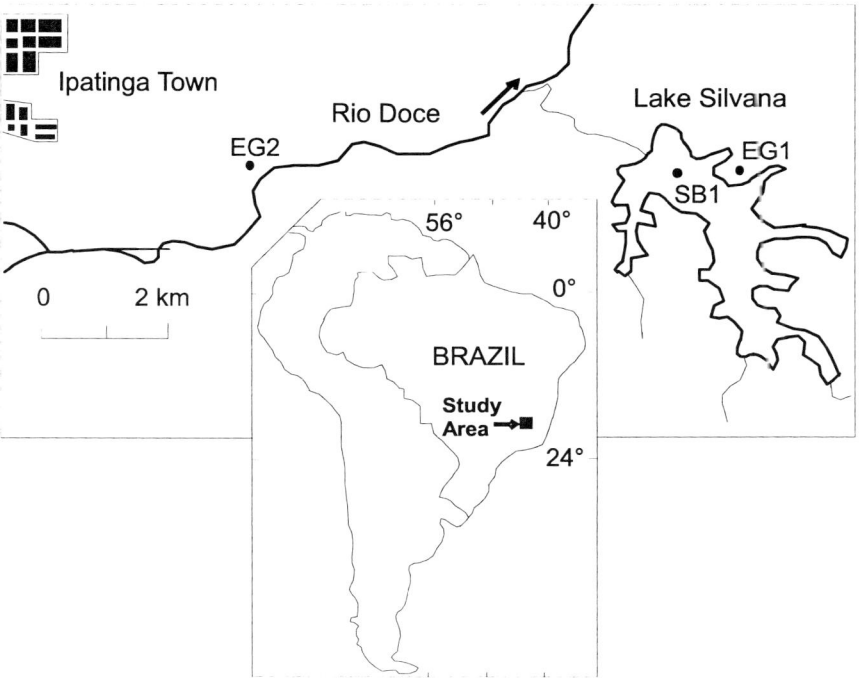

Figure 2.2 Location of the study area and sampling sites

2.1.1 Geology and geomorphology

Most of the southeastern region of Brazil exposes the Atlantic continental shield, which constitutes the Archean basement throughout the Atlantic coast. Archean terrains were reworked during the tectono-magmatic cycles named *Transamazônico* (~2000 Ma) and *Brasiliano* (~600 Ma), giving rise to both the migmatitic complexes of Minas Gerais and Bahia (DNPM, 1984). In its middle course, Rio Doce drains a deeply weathered sequence of Archean/Late Proterozoic gneisses and granites with relicts of mafic rocks, corresponding to the gneissic-migmatitic complex of Minas Gerais (Fig. 2.3). This unit constitutes the basement of the Rio das Velhas and Minas supergroups in the *Iron Quadrangle* [*Quadrilátero Ferrífero* (DNPM, 1984)]. Biotite-rich gneisses with lenses of amphibolite, named Piedade gneiss, form the bedrock of the study area. They were probably derived from the gneissic-migmatitic complex in the Late Proterozoic (IBGE, in press). The Piedade gneiss consists of banded biotite gneiss with quartz veins, pegmatites and amphibolitic lenses. Banding ranges from some millimeters to some centimeters, with alternating felsic and mafic bands. This litho-structural unit is believed to have formed between 2400 and 2250 Ma, according to Rb-Sr dating (IBGE, in press).

The lateritic mantle developed over these rocks may reach a thickness of 30 m. Prevailing kaolinite-producing weathering of the Precambrian basement is illustrated by the clay mineral distribution at the Atlantic shelf area off the Rio Doce, with an average kaolinite content of 83% (Tintelnot, 1995).

The study area lies in a physiographic unit named *Interplateau Lowlands of the middle Rio Doce Valley*, which is confined between escarpments bordering the Southeastern Brazilian Plateau (Meis and Monteiro, 1979). It consists of an elongated depression with altitudes ranging from 200 to 400 m a.s.l., likely controlled by structural trends of the rocky substratum that strike NE-SW. The hilly uplands of the Southeastern Brazilian Plateau consist of smooth, undulating surfaces of low relief carved in the Precambrian bedrock. The rolling convex-concave slopes and slightly inclined valley bottoms are typical landscapes of humid regions (Meis and Moura, 1984). Meis and Monteiro (1979) investigated the morphostratigraphic evolution in the middle Rio Doce Valley and recognized colluvial deposits interfingered with late Pleistocene and Holocene alluvial deposits, forming the so-called *rampa de colúvio* deposits. The formation of these deposits is attributed to accelerated rates of slope retreat in areas of increased humidity, generating gently inclined, slightly concave flat surfaces. Modenesi (1988) described a series of superposed colluvial deposits in the Campos do Jordão Plateau with more weathered materials at the base and less altered sediments at the top and buried humic horizons, which gives further evidence of late Quaternary mass movements in southeastern Brazil.

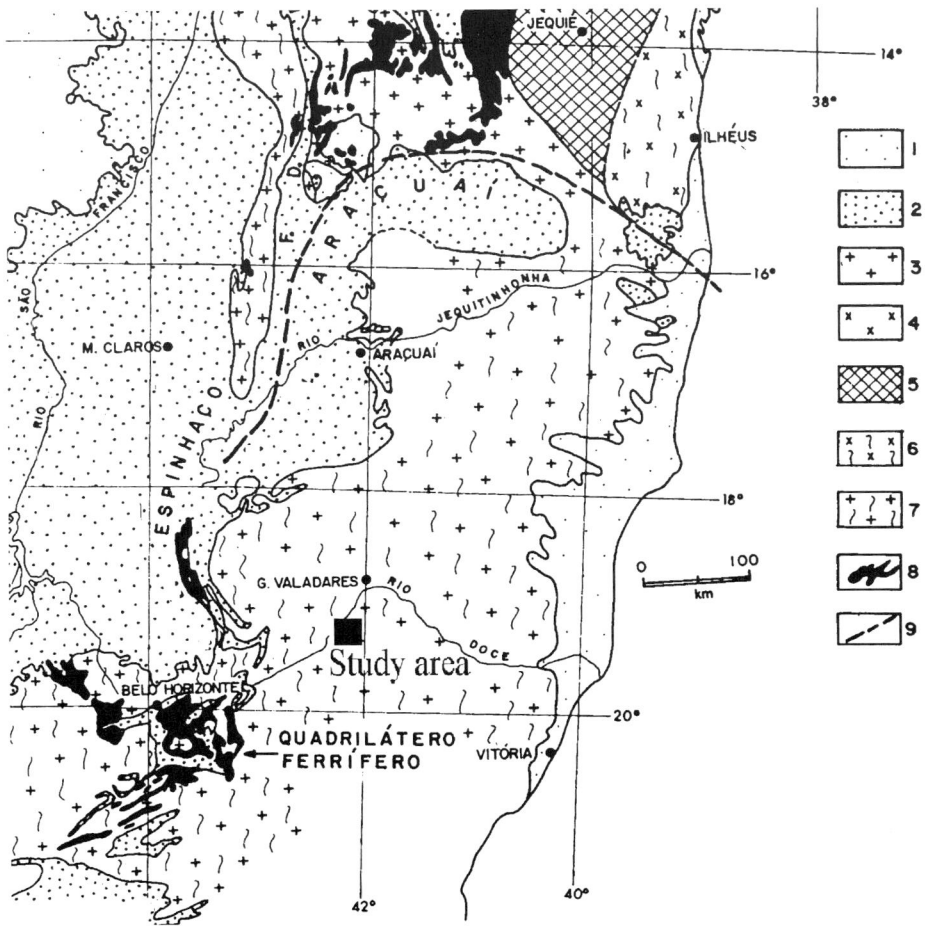

Figure 2.3 Geological map of a part of southeastern Brazil. (1) Phanerozoic
sedimentary rocks; (2) Proterozoic orogenic belts; (3) granite-greenstone
terrains; (4) granite-greenstone terrains reworked in the Transamazonic
cycle; (5) granulitic terrains; (6) granulitic terrains reworked in the
Transamazonic cycle; (7) migmatite-granulitic and granite-greenstone
terrains reworked in the Transamazonic and Brazilian cycles; (8) greenstone
belts; (9) Boundary between the *São Francisco* craton (north) and the
Araçuaí and *Ribeira* orogenic belts (modified after DNPM, 1984).

According to Bigarella and Andrade-Lima (1982), the landform evolution in Brazil has a polycyclical character related to the action of two main sets of morphoclimatic processes. A lateral degradation process is attributed to prolonged aridity during the Pleistocene, while valley incision is thought to take place during interglacial intervals of humidity. During climatic changes from humid toward semiarid conditions, the weathering mantle is believed to move downslope clogging the basin's outlet and therefore filling up the valleys with sediments. These authors observed that the Brazilian Holocene terraces consist usually of silty-clayey sediments and are thought to represent climatic fluctuations with forest retreat favoring slope erosion.

2.1.2 Climate and vegetation

The middle Rio Doce Valley lies in a region under particular climatic constraints. It is situated in the transition between contrasting climatic and vegetation domains, both from east to west and from north to south (Fig. 2.4). Therefore, it is thought to consist of a region potentially indicative of Holocene climatic changes. Southeastern Brazil has a marked climatic diversity ranging from warm and humid tropical climate, without or with a dry season restricted to 1-2 months, to semi-arid conditions, with a dry season of 6 months (Nimer, 1989). The vegetation types range from Atlantic rain forest in moister regions to thorn scrub savanna (*caatinga*) in drier ones. Semideciduous forest and wooded savanna (*cerrado*) correspond to intermediate levels of humidity. The transition from tropical Atlantic forest to semideciduous forest and then to savanna formations is from east to west, while *caatinga* occurs only in the northern part of SE Brazil (Fig. 2.4). Semideciduous forest occurs in regions with annual precipitation between 1000 and 1500 mm, mean annual temperature between 18 and 22 °C and a dry season of 3 to 5 months (IBGE, in press; Nimer, 1989). *Cerrado* formation occurs in regions with mean annual precipitation between 1000 and 1750 mm, mean annual temperature between 20 and 26 °C and a dry season of 5 to 6 months. The xenomorphic *cerrado* vegetation is well adapted to fire, as indicated by the occurrence of the thick corky bark of several woody species (Coutinho, 1982). Semideciduous forest represent the primary vegetation in the middle Rio Doce Valley, where the mean annual precipitation and the mean annual temperature reach 1250 mm and 22 °C, with 4 to 5 dry months. The most relevant occurrence of semideciduous forest lies in the Rio Doce Forest Park, between the townships Timóteo and Marliéria. The large-scale production of charcoal for local steel industries has led to a progressive destruction of the primary vegetation during the last decades. The secondary vegetation that grows after deforestation is characterized by the low diversity of species (IBGE, in press).

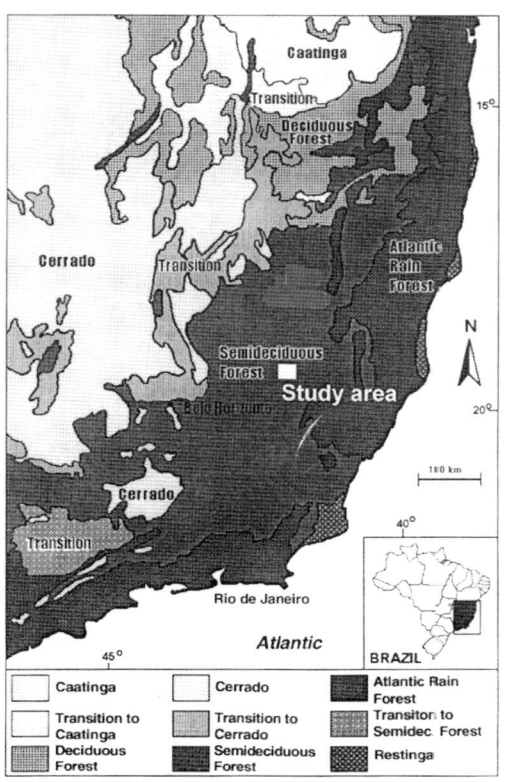

Figure 2.4 Vegetation map of southeastern Brazil (IBGE, 1993, after Behling, 1995).

2.2 The dammed-valley lakes of Rio Doce

Lake Silvana is one of almost forty closed-basin lakes occurring in the middle course of Rio Doce. Several lakes are surrounded by Atlantic Rain Forest and semideciduous forest, where the primary vegetation was preserved. Primary vegetation occurs mainly in the Rio Doce Forest Park. A large area surrounding the Forest Park, which comprises a part of the Lake Silvana catchment, is covered by secondary forest with small trees and *Eucalyptus*, as the charcoal produced from *Eucalyptus* wood is largely used by the steel industries for the ore smelting. A satellite image (Spot) of the study area gives a picture of some dammed-valley lakes on the right bank of Rio Doce, including Lake Silvana, and the preserved primary vegetation restricted to the Rio Doce Forest Park on the left river bank (Fig. 2.5).

Pflug (1969) has recognized 136 former and current lake basins stretching for 9 km along both banks of the Rio Doce. Among them, only 38 are still filled with water, indicating that the lake system is undergoing a process of desiccation. The smaller, former lake basins have dried up. The lake levels lie about 20 m above the present level of the Rio Doce and the lake sizes range from 0.1 to 5 km^2. The lakes are relatively shallow, with maximum depths ranging from 6 to 30 m.

Sedimentological and geomorphological studies were carried out in the middle Rio Doce Valley by Pflug (1969) and Meis and Monteiro (1979), who focused on the origin of the lake system and the morphostratigraphic evolution of the basin during Upper Quaternary times, respectively. According to these authors, the lake system originated from the accumulation of coarse clastics as an anastomosing channel system, which dammed the tributary valleys through differential deposition rates in the main valleys relative to their tributaries. According to Meis and Monteiro (1979), this alluvial sequence may reach a thickness of 35 meters.

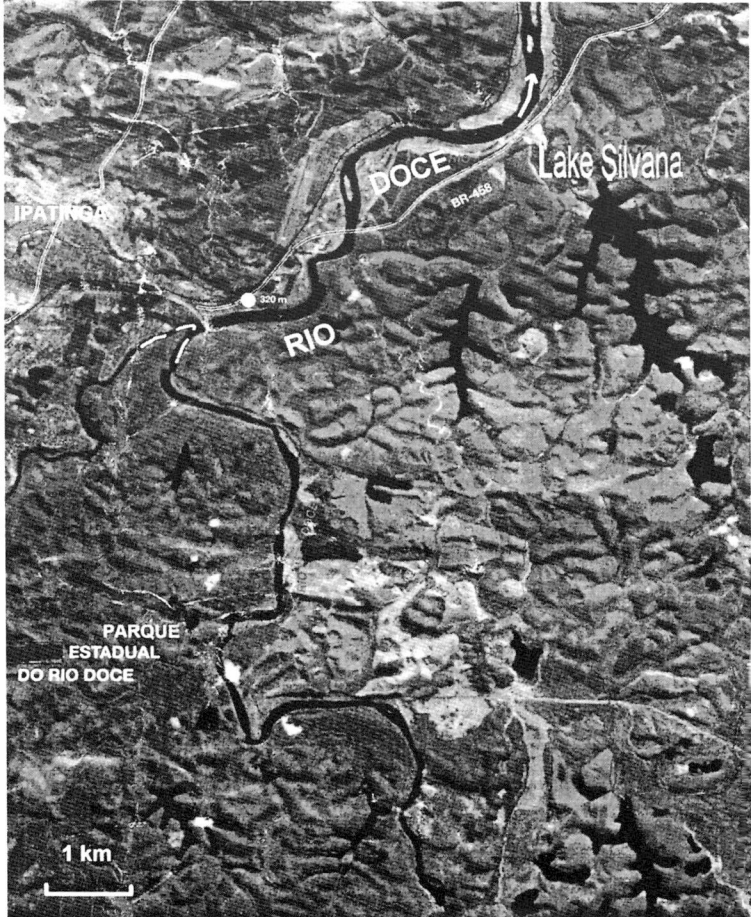

Figure 2.5 Satellite image (Spot) of the study area.

Pflug (1969) attributes the formation of the Rio Doce lakes to a polycyclic development of the river valley. Rio Doce flows through a wide plain which is believed to have formed during a semi-humid period that yielded a 15 to 20 km wide valley entrenched into the Precambrian basement. After this pediplanization period, the river system is thought to have undergone a pronounced dissection, probably due to a lower base level of erosion that yielded a lowering of the river bed to about 70 m below the pediplain. According to Pflug (1969), the dissection phase has led to the formation of a dendritic system of tributary valleys. This erosional phase is believed to have been followed by a period of aggradation of the river bed that caused the accumulation of the above-mentioned alluvial terrace (Fig. 2.6). The author pointed out that the material of the alluvial terrace is likely to have been deposited under a semiarid climate, since it contains fresh feldspars and micas, indicating prevailing physical weathering. Pflug (1969) suggests that the formation of the Rio Doce lakes has occurred due to a geomorphic response to prevailing semiarid conditions in the late Pleistocene, as indicated by a preliminary radiocarbon dating from carbonaceous matter of the alluvial terrace (14,160 ± 500 yr B.P.). However, due to the coarseness of the dated material, the possibility of sediment reworking should be considered.

Figure 2.6 Lake Silvana and Rio Doce. Note the alluvial terrace on the right river bank.

Meis and Monteiro (1979) observed that the depositional sequence filling the lower order tributaries shows alluvial fans partially derived from the re-working of slope-wash sediments, as indicated by interfingered alluvial/colluvial sediments, the so-called *rampas*. The authors recognized a series of step-like benches that reflect the recurrence of processes during the Quaternary, characterized by accelerated rates of slope retreat in areas of increased humidity. It has been observed that the displacement of regolith debris by mass wasting processes results in structureless, poorly-sorted sediments. The denudational activity over the hillslopes is thought to be closely linked to variations in the paleohydrological regime. According to these authors, there has been a general trend toward valley incision under wetter climates during post-glacial times, as suggested by the existence of relatively few areas where aggradation of valleys occurred in the Pleistocene-Holocene boundary. The aggradation phase responsible for the damming of tributary valleys in the middle Rio Doce is believed to be one of these few examples where an increase in moisture leads to valley aggradation rather than valley incision. This interpretation has been based on the properties of the sandy terrace, several meters thick, where lenses of poorly-sorted fine particles indicate the occurrence of mud flows and intense activity of flowing water. Meis (1977) describes a sedimentary sequence in a shallow lake near Ipatinga, where the weathered Precambrian basement is covered by silty alluvium overlain by lacustrine clays and silty clays with sandy lenses. According to this author, the natural dams that gave rise to the lake system formed by watershed destruction through colluviation, coupled with fluviatile aggradation. The occurrence of sandy lenses in lacustrine silty clays has been attributed either to more intense flows during increasing rainfall or a Holocene lacustrine retreat. Meis (1977) revised Pflug's data and recognized the following morphostratigraphic sequence in the middle Rio Doce Valley:

- Middle Pleistocene: a period of valley incision was followed by the formation of thick sequences of poorly sorted slope deposits in colluvial fans;

- Late Pleistocene: a period of river aggradation and reworking of the colluvial fans was associated with the deposition of a thick alluvial sequence along the main rivers;

- Pleistocene-Holocene boundary and early Holocene: Rio Doce and its main tributaries degraded their channels and a new drainage net formed. A further terrace formed through the rapid retreat of the soft, sandy banks of the newly developed channels;

- Middle to early Holocene: the main rivers continued their degradation work which was interrupted by one phase of terrace formation.

Costa and Schuster (1988) carried out a world-wide survey on the numerous kinds of dams that form by natural processes, including volcanic, landslide, glacial and fluviatile dams. The most common constraints that lead to dam-forming landslides are excessive rainfall and snowmelt and earthquakes. The types of mass movements that form landslide dams are rock and debris avalanches; rock and soil slumps and slides; and mud, debris and earth flows. A significant percentage of landslide dams consists of mud, debris and earth flows. Most of these dams have been caused by relatively high-velocity debris flows derived from tributary valleys. The authors observed that dams formed in this manner are not high, and if composed of noncohesive material, they commonly overtop soon and breach rapidly. In the case of slower debris flows and/or flows of cohesive materials, longer-lived dams may form.

2.3 The lacustrine environment in the Rio Doce lakes and related redox processes

Lacustrine environments located in areas either with abundant vegetation or with nutrient-rich waters that favor the growth of aquatic plants and plankton are commonly characterized by large amounts of organic matter on the bottom, leading to prevailing reducing conditions in the hypolimnion through organic decay. The bottom of these lakes tends to present organic gels (*gyttia*) that are characterized by low density and organic matter contents ranging from 10 to 50% (Rose et al., 1979). Organic gels may form sections up to 10 m thick with a typical accumulation rate between 40 and 60 cm/1000 yr.

A comparative study of 15 lakes (Tundisi et al., 1997) provided the main limnological features of the Rio Doce lake system. Most lakes show weak or no stratification during the winter, whereas lakes deeper than 10 meters or sheltered from wind action present a classical summer stratification, with chlorophyll-a accumulation in the metalimnion and an elevated concentration of ammonium (NH_4^+) in the anoxic hypolimnion. Anoxic bottom waters also show increasing values of conductivity, alkalinity, bicarbonate (HCO_3^-), and slightly lower pH. Dissolved oxygen concentrations in euphotic layers range from 5 to 8 mg/l (Mitamura and Hino, 1997). Even surface waters are unsaturated with dissolved oxygen, suggesting that the respiration rate exceeds the rate of primary production.

The natural eutrophication of many lakes in this system is thought to be a result of the accumulation of terrestrial organic matter on the lake bottoms, which leads to oxygen depletion during stratified periods (Tundisi et al., 1997). The main allochthonous origin of organic sediments has been confirmed by investigating the source of sedimentary organic matter in four lakes (Nishimura et al., 1997). Biological markers of fatty acids and sterols were used for tracing back land- and plankton-derived organic carbon in sediments. The minimum proportions of land-derived organic carbon, between 66 and 74%, were found in surface sediments, while in deeeper layers the ratio of allogenic/authigenic organic carbon increases up to 93%. This has been explained with a preferential biodegradation of planktonic relative to terrigenous matter during early diagenesis, rather than with changes in relative organic inputs from both sources.

According to Meis and Tundisi (1997), the lake typology proposed for temperate lakes on the basis of trophic state, chemical composition of sediments, primary production and nutrient concentration may be misinterpreted when used for tropical lakes, where short-term cycles – hourly or diurnal – occur. Therefore, the authors suggest, based on the study of 12 lakes of the middle Rio Doce Valley, that a lake typology for tropical lakes should also consider long-term changes such as the geomophological evolution of the lake basin and its morphometric features. It has been demonstrated that several limnological features are strongly dependent on the morphometric characteristics of the lakes. The morphometry and the geographic placement of the lakes present strong influence on wind action, and consequently on thermal and chemical stratification.

Sediment geochemistry, authigenic mineralogy and mineral transformations during early diagenesis in the lakes of Rio Doce are almost unknown. Saijo et al. (1991) determined the vertical distribution of organic C, N and P in surficial sediments of four lakes. It has been observed that the water content was very high, ranging from 96% at the surface to 91% at a depth of 50 cm. Organic carbon and nitrogen contents in sediments of Lake Carioca (20% and 2.3%, respectively) were higher than those of Lake Dom Helvecio (12% and 1.1%). Phosphorus content showed minor variations between both lakes, with mean values of 0.1%. These values are approximately in agreement with the commonly assumed composition of organic matter, following the proportions 106:16:1 for C:N:P (Stumm and Morgan, 1996).

Berner (1981) pointed out that the classical approach of characterizing sedimentary environments in terms of Eh-pH diagrams has a limited practical use for geochemical studies of modern sediments. In fact, pH values do not vary significantly in most aquatic systems and other hydrochemical parameters, which are involved in sedimentary redox reactions, such as alkalinity, SO_4^{2-}, NO_3^-, N_2, NH_4^+, HCO_3^- and CH_4, are not detectable by measurements of redox potential (Eh). The author proposes, therefore, a new geochemical classification based on the concentration of dissolved oxygen and dissolved sulfide (H_2S plus HS^-), dividing the sedimentary environment into oxic, post-oxic, sulfidic and methanic. From this classification, Fe and Mn minerals can be used to characterize each sedimentary environment (Table 2.1).

Table 2.1 Geochemical classification of sedimentary environments after Berner (1981)

Environment	Characteristic phases
Oxic ($O_2 > 10^{-6}$ M)	hematite, goethite, MnO_2-type minerals; no organic matter
Anoxic sulfidic ($O_2 < 10^{-6}$ M, $H_2S > 10^{-6}$ M)	pyrite, marcasite, rhodochrosite, alabandite; organic matter
Anoxic nonsulfidic post-oxic ($H_2S < 10^{-6}$ M)	glauconite, siderite, vivianite, rhodochrosite; no sulfide mineral; minor organic matter
Anoxic nonsulfidic methanic	siderite, vivianite, rhodochrosite; earlier formed sulfide minerals; organic matter

The author predicts the occurrence of minor organic matter in the anoxic-nonsulfidic-post-oxic environment assuming that such conditions can be reached only if the amount of decomposable organic matter is not sufficient to bring about sulfidic conditions. However, the formation of the same mineralogical assemblage seems also predictable for environments with abundant organic matter and low-sulfate waters.

Due to the Fe- and Mn-rich, highly-leached lateritic cover that covers large areas in tropical regions, iron becomes the most available metal for geochemical interactions between the dissolved and solid phase in tropical aquatic systems, since Na, Ca, Mg and K are normally depleted in the catchment soils, and Si and Al present a low mobility. The redox cycling of Fe and Mn has pronounced effects on the adsorption of trace elements and their fluxes under different redox conditions (Song and Müller, 1999; Balistrieri et al., 1994; Gunkel and Sztraka, 1986). Stumm and Morgan (1996) emphasized that the reduction of manganese occurs at higher redox potentials than does the reduction of iron, whereas the reoxidation of iron occurs faster.

Since the bulk of the solids dissolved in water depends primarily on the composition of catchment rocks, freshwater systems surrounded by rocks with low sulfide contents have commonly very low sulfate concentrations. Furthermore, lake waters in igneous and metamorphic terrain commonly have lower dissolved solids content than lakes in sedimentary terrain, which tend to present high nutrient contents (Rose et al., 1979).

Alkalinity of lake waters may be indicative of redox processes and decomposition of organic matter. Alkalinity is normally produced by increasing concentrations of cations and by processes removing strong acid anions. In temperate lakes, the major processes that commonly generate alkalinity are: biological reduction of SO_4^{2-}, exchange of H^+ for Ca^{2+} in sediments, and biological reduction of NO_3^- (Schindler et al., 1986). For tropical lakes with low-sulfate waters, however, alkalinity seems to increase in anoxic waters as a result of the diffusion of Fe^{2+} and Mn^{2+} from sediments and the reduction of NO_3^- to NH_4^+. Schindler et al. (1986) presented the equation that normally defines alkalinity in temperate lakes:

$$Alk = (Ca^{2+} + Mg^{2+} + Na^+ + K^+ + NH_4^+) - (SO_4^{2-} + NO_3^- + Cl^-)$$
$$= HCO_3^- + OH^- + CO_3^{2-} + Org^- - H^+$$

Large amounts of Fe and Mn are normally transported into tropical lakes, either in solution from the reduction in the humic layer of lateritic terrains or as Fe-Mn oxides, which tend to be reduced in anoxic bottom waters. Dissolved Fe and Mn species may (re)precipitate on the bottom as oxyhydroxides, carbonates, sulfides and/or metallo-organic compounds, depending on varying physical and hydrochemical parameters and biological activity. Stumm and Morgan (1996) stressed that most redox processes involved in the cycling of metals in natural waters are microbially mediated, as nonphotosynthetic organisms tend to restore equilibrium by catalytically decomposing the unstable products of protosynthesis through energy-yielding redox reactions. According to these authors, a typical sequence of redox reactions in aqueous systems is characterized first by the reduction of O_2 (aerobic respiration), followed by the reduction of NO_3^- to N_2 (denitrification), MnO_2, NO_3^- to NH_4^+ (nitrate reduction), FeOOH, organic matter (fermentation), SO_4^{2-} and CO_2 (methane fermentation). The authors observed that in aqueous systems containing excess organic matter, such as an eutrophic lake or a polluted system, fermentation reactions may occur prior to the reduction of FeOOH.

2.4 Chemical weathering in tropical regions

The Brazilian and African shields were submitted to humid tropical climates since the late Jurassic, when the opening of the Atlantic ocean was accompanied by a northward movement of Africa and South America, leading to the development of favorable climatic conditions for laterite formation (Parrish et al., 1982; Tardy et al., 1988). Consequently, both continents are covered by a thick lateritic cover, where bauxites are largely distributed. In Brazil and East Africa, bauxites are generally massive, without pisolitic structure and mainly formed by gibbsite, while boehmite is almost absent (Tardy et al., 1988; Valeton, 1972). Valeton (1972) summarizes the main phases of bauxite formation in the earth's history. There is a clear predominance of boehmite and diaspore in deposits formed during Mesozoic and Palaeozoic times, while those formed in the Tertiary are generally gibbsitic. Bauxite deposits in Brazil and East Africa are included in the last group.

Ferricretes, or iron crusts, seem to be better preserved under seasonally contrasted tropical climates with a long dry season, as in humid regions they tend to break down (Nahon, 1986; Tardy and Roquin, 1992). Seasonally contrasted tropical climates are thought to be associated with the formation of iron crusts because alternating reducing and oxidizing conditions are required to allow successive small-scale redistribution of iron by leaching (Boulet, 1978 in Nahon, 1986). The formation of thick iron crusts may require more than 1 Ma, and certainly requires less than 6 Ma, depending on the nature of the parent rock, the intensity of climatic changes, the tectonic evolution, and other factors (Nahon, 1986). This author indicated that all these factors being considered, it is likely that most of the ferricrete profiles in West Africa began to form at the end of the Tertiary. Similarly, McFarlane (1983) estimates a period required for the formation of lateritic profiles of 1-10 Ma.

The term laterite has a broad usage and includes bauxites, ferricretes, iron and aluminium duricrusts, mottled horizons, pisolite − or nodule − bearing materials and is extended to the formations or horizons that are parts of red or yellow ferralitic soils, tropical ferruginous soils and other kaolinitic formaticns (Tardy, 1992). Chemical weathering in humid tropical regions leads to the formation of typical mineral assemblages made up of secondary minerals, commonly represented by kaolinite, gibbsite, hematite and goethite and residual quartz, while anatase and zircon are common trace constituents (Irion, 1984; Lucas et al., 1993; Curi and Franzmeier, 1984; Kronberg et al., 1979). Dissolution of soluble minerals and geochemical reactions within the pedologic mantle is mainly controlled by hydrolysis, which in turn is climate-dependent (Cheswo-th, 1992). Microbiological activity in topsoils appears to be a further relevant factor controlling the surficial accumulation of silicon in equatorial ecosystems (Lucas et al., 1993). Merino et al. (1993) have observed that during lateritic weathering, pseudomorphic replacement of parent-rock silicates by kaolinite and Al and Fe oxides is systematically accompanied by dissolution voids in overlying parent minerals.

A typical lateritic profile can be schematically represented by three zones of major horizons, namely, zone of alteration at the base; nodular or mottle zone in the middle part of the profile; and an upper, soft zone (Braucher et al., 1998; Tardy, 1992). The zone of alteration consists of coarse and fine saprolite, where the structure of the parent rock is roughly preserved. Fine saprolite is characterized by a complete or partial alteration of most of the primary minerals, while iron and aluminum remain almost immobile. Near the groundwater table, goethite tends to predominate over hematite, and gibbsite over kaolinite, due to the high degree of hydrolysis (Tardy and Roquin, 1992). The nodular or mottle zone generally shows accumulations of iron oxyhydroxides and kaolinite (mottles) contrasting with bleached domains. In general, hematite predominates over goethite in the zone of nodule formation due to dehydration. Bleached domains consist mainly of quartz and kaolinite and exhibit a white or gray color due to the de-ferruginization of the previously associated kaolinite and iron oxyhydroxides (Nahon, 1986). The soft zone is characterized by a relative accumulation of quartz, kaolinite and iron oxyhydroxides, either resulting from the dissolution and dismantling of glaebular material. The mechanism of dismantling is thought to be related to the rehydration of hematite and kaolinite resulting in the formation of goethite and gibbsite at the top of the profiles (Nahon, 1986; Tardy, 1992).

Bioclimatic changes may lead to significant transformations of pedologic systems. Transformation of original sequences is expressed by disorganization of upper horizons that are exposed to such changes. Nahon (1991 and references therein), based on studies of pedologic mantles in Senegal, have suggested that under drier climatic conditions, microaggregates of about 100 µm in diameter composed of kaolinite and iron oxyhydroxides undergo a process of segregation corresponding to a structural disorganization without major mineralogical changes. This segregation is thought to be responsible for the transition from red soils to beige soils in the center of plateaus.

The formation of beige horizons at the top of weathering profiles has been recognized in several sites of the Amazonian Lowland (Irion, 1984), where beige soils are generally composed of poorly ordered kaolinite, quartz, goethite and gibbsite. This mineralogical assemblage seems characteristic of the most advanced weathering stage in tropical regions. The predominance of goethite is commonly associated with beige soils (Curi and Franzmeier, 1984; Irion, 1984). Gibbsite, however, may occur either in red ferralitic soils (Curi and Franzmeier, 1984) or in beige soils (Irion, 1984).

3 Methods

3.1 Sampling

One sediment core and two soil profiles were collected in Lake Silvana and its catchment. Core SB1 is 12.7 m long and was collected using a piston corer capable to reach great depths with excellent recovery rates. This has been assessed through the length of each core section after extrusion, which indicated negligible loss of material.

A platform was built over two small boats and fixed with three 12-m-long *Eucalyptus* trunks (Fig. 3.1). A total of 72 sub-samples were taken at 10- or 20-cm-intervals. One-meter-long core sections were carefully extruded and sealed in PVC tubes – with the same diameter as the core – for transportation to the Institute of Environmental Geochemistry at the University of Heidelberg, Germany.

Figure 3.1 Coring platform - core SB1.

Two profiles of the weathering mantle (EG1 and EG2) were sampled, providing 21 samples, where road cross-sections enabled the exposure of the entire pedologic cover – for the profile EG1, however, only the first 3.5 meters could be sampled. Sampling points within the weathering profile, which may reach a thickness of 30 meters, were selected according to changes in macroscopic features, such as color, texture and pedological and lithological structure

Water samples were collected at various depths above the core site SB1 through a small pump and stored in polyethylene bottles. Three samples from selected depths – surface, intermediate and bottom waters – were acidified with HNO_3 to pH 1 for determination of dissolved Fe and Mn, after filtering through cellulose membrane with pore size of ca. 0.45 µm. This preservation procedure is indicated to avoid losses of dissolved metals due to their adsorption on suspended sediments, on precipitates and on the container walls (Fletcher, 1980). The possible effect of increasing apparent dissolved metal concentrations as a result of the increased acidity, could be estimated from the total suspended solids (5–20 mg/l) and the average Fe and Mn concentration in surface sediments – 10% and 600 µg/g, respectively. Considering an unlikely total metal extraction from suspended sediments during storage, even so the contribution from the suspended load to the measured concentration of dissolved Fe and Mn would not be higher than 20% and 15%, respectively. Immediately after collection, *in situ* measurements of pH, Eh, electrical conductivity, temperature and dissolved oxygen were conducted.

3.2 Analyses

A total of three ^{13}C-corrected radiocarbon age determinations with accelerator mass spectrometry (AMS) were carried out from samples of plant debris by Geochron Laboratories (Krueger Enterprises, Inc.). Plant fragments were separated from any sand, silt, or other foreign matter. Samples were then treated with hot diluted HCl to remove any carbonates. After washing and drying, the cleaned charcoal was combusted and the carbon dioxide was recovered for analysis (Geochron report GX-23912).

Granulometric distribution within the core was obtained by wet sieving, comprising the following grain size fractions: <20 µm, 20–63 µm and >63 µm. Nylon sieves and distilled-deionizated water were used to avoid contamination.

Measurements of heavy metal concentrations in sediments were performed in the <20 µm fraction. All samples were dried at 30°C until constant weight, and digested using *aqua regia* (HNO_3/HCl, 1:3), for 3 hours at 160°C (Müller, 1979;

Song and Müller, 1999). Iron, Al, Mn, Pb, Zn, Cu, Ni and Cr concentrations were determined by flame atomic absorption spectrometry – AAS (Perkin Elmer 4100), whereas Hg was measured by cold vapor AAS (Thermo Separation Products 3200). In addition, concentrations of Fe, Al, Pb, Zn, Cu, Ni and Cr were measured in selected samples using both flame AAS and X-ray fluorescence – XRF (Siemens SRS 300) for assessing the efficiency of the *aqua-regia* extraction. Si, Ti and K concentrations were achieved only by XRF. For XRF analyses, samples were fused to give a homogeneous glass disc. This technique of sample preparation reduces variations in the mass absorption coefficient and allows more accurate analyses (Fletcher, 1980). Thirty-one of the 93 samples of sediments and soils were prepared in duplicate for AAS analyses: the measurements were in close agreement – ± 15% (Appendices A1–A7). Accuracy control was performed by including a calibrated standard sample in each sample set.

Although *aqua-regia*-soluble concentrations do not correspond to total metal concentrations, the relative vertical distribution of metal concentrations in both the sediment core and the soil profile EG2 gave similar patterns when comparing the results from both analytical methods, AAS and XRF. The similar distribution patterns resulted from less variable AAS/XRF ratios for each metal with depth in the sediment core, indicating a relatively homogeneous mineral matrix. In the soil profile, AAS/XRF ratios show more accentuated variations, probably due to a less homogeneous mineral matrix. Despite of these variations within the soil profile, the large geochemical contrasts between horizons largely compensate the low efficiency of *aqua-regia* extraction for certain horizons (Appendices A1–A7).

Mineralogical analyses of sediments and soils were performed from dry samples of the fraction <20 µm using an X-ray diffractometer – XRD (Siemens 500), and a scanning electron microscope – SEM (Leo 440) with an energy dispersive spectroscope – EDS (Oxford 7060). XRD analyses were performed in randomly oriented powder samples under the following operational conditions: Cu-Kα radiation, 40 kV and 25 mA. X-ray diffractograms were interpreted using the software Siemens Diffrac AT V.3.2. For the SEM/EDS analyses, samples were sputtered with gold. Optical-microscopy examinations required sample preparation with a binding agent (poly-vinyl alcohol) for slicing of thin sections. Here, the sediment fractions <20 µm and 20–63 µm were examined in order to identify major mineral constituents. Contents of carbonate minerals were calculated by measuring the CO_2 pressure produced after adding heated HCl. Since carbonate contents are proportional to the generated CO_2 pressure, quantitative determinations ($\pm$ 1%) are obtained through standard calibration (Müller and Gastner, 1971).

Diffractograms from XRD together with measurements of major elements from XRF were used for the quantitative determination of major minerals. Stoichiometric calculation was based on the assumption that muscovite content is a function of K concentration, and contents of Fe-oxide minerals are proportional to Fe concentration excluding siderite-Fe, so that there is only one content of each mineral, kaolinite, gibbsite and quartz for a given concentration of Si and Al. Goethite/hematite ratios were estimated from the height and width of the differential peaks.

Concentrations of total C and S were measured by heating the solid sample to 1800 °C in a high-frequency induction oven coupled with an infrared absorption spectrometer (Leco CS-225). Concentrations of organic C could be calculated from the measured concentrations of total C and inorganic C, since siderite is the only carbonate mineral present in the samples.

Pollen analyses were carried out by Dr. Hermann Behling from the Center for Tropical Marine Ecology, Bremen, Germany. Samples of 1 cm^3 were taken at 20 cm intervals (in the lower core section intervals are shorter) along the profile. Samples were prepared using the standard treatment with sodium pyrophosphate and heavy liquid separation by bromoform (Faegri and Iversen, 1989). Pollen preparation included addition of exotic Lycopodium spores to determine the

4 Results

4.1 Lake sediments - paleoenvironmental proxies from core SB1

Four major paleoenvironmental zones are represented by distinct sedimentary sequences within the core SB1. Different phases of the hydrological cycle in Lake Silvana were inferred from contrasting geochemical and mineralogical patterns, which are consistent with changes in vegetation indicated by pollen data since the early Holocene. XRD analyses in the <20 µm fraction of samples collected at regular core intervals revealed a dominance of poorly crystallized kaolinite, followed by goethite, muscovite and quartz. Varying contents of gibbsite and siderite occur since the transition from zone I to zone II.

4.1.1 Paleoenvironmental zone I (1270-1180 cm)

The lowermost section of core SB1 differs from the overlying sections firstly in presenting no laminated structure, besides the occurence of abundant plant rootlets and carbonized plant debris (Fig. 4.1). Accelerator mass spectrometer (AMS) radiocarbon dating from the upper and the lower segments of this zone gave ages of 9430 ± 60 and 10,120 ± 50 yr. B.P., respectively. Similarly, Ybert et al. (1996) have determined ages between 9500 and 9000 years B.P. for the lowermost zone of a 7.4-m-long sediment core from Lake Dom Helvecio, situated ca. 30 km to the south.

In zone I, yellowish silty sediments are characterized by small amounts of sand (8%). Through thin sections of the sand fraction, the dominance of Fe-kaolinite as pseudomorphs after biotite were observed. The silty-clayey fraction – <20 µm (57%) and 20–63 µm (35%) – revealed a grain-size distribution that characterizes the poor sorting of this unit. XRD analyses and stoichiometric calculation from measurements of major elements in the <20 µm fraction – Si, Al, Fe, Ti and K – together gave kaolinite (80%) and goethite (15%) as major constituents, supporting the observations by optical microscopy (Figs 4.2, 4.3).

Low contents of organic matter occur in the fine structureless matrix, whereas plant debris is concentrated at the base of this unit. Analyses of major and trace elements in the <20 µm fraction gave relatively high concentrations of Fe, Ti, Cr and Cu, contrasting with the overlying core sections. On the contrary, gibbsite and Hg contrast with the overlying sections due to their lower concentrations (Figs. 4.3, 4.5). An oxic nature and a detrital origin of zone I is indicated by its physical properties, such as color, sorting degree and absent lamination, and by the relatively high contents of likely detrital goethite, as well as by the likewise elevated concentrations of Ti, Cr, Pb and Cu (Figs. 4.2, 4.4).

Pollen samples are characterized by 95% of herbaceous pollen (*Poaceae* and *Cyperaceae*) indicating the predominance of grassland vegetation with probably very small gallery forests along the waterways. This paleovegetation indicates that sediments were deposited under a markedly colder and drier climate than the present one (Fig. 4.5).

Figure 4.1 The lowermost section of core SB1 (1270 cm) showing the radiocarbon-dated plant debris and rootlets (core-section diameter = 4 cm).

4.1.2 Paleoenvironmental zone II (1180-1160 and 990-970 cm)

In zone II, grain-size distribution is characterized by a sharp increase in the <20 µm fraction, up to 93%, with 7% of the 20–63 µm fraction and no sandy fraction. The well-sorted clayey sediments are finely laminated and show a reddish-brown color and slightly elevated contents of organic C, up to 0.93%. Samples of the <20 µm fraction are composed of kaolinite (60%), gibbsite (10%), goethite (10%), siderite (8%), muscovite (5%) and quartz (5%). It is interesting to note the occurrence of siderite and gibbsite contrasting with zone I (Figs. 4.2, 4.3).

Trace element analyses of the <20 µm fraction gave elevated concentrations of Mn and Hg, and reduced Cr, Pb and Cu relative to zone I (Figs. 4.4, 4.5). In zone II, both the geochemical pattern and the occurrence of siderite suggest the onset of anoxic conditions and the formation of a shallow lake shortly after 9430 yr B.P. The finely laminated and well-sorted sediments also indicate an environment of lacustrine sedimentation. Moreover, either a low water level or a low biological productivity, or both, is suggested by the low organic C contents.

Zone II is characterized by 50-70% of herbaceous pollen, 20-30% of tropical shrubs and trees and 5-10% of arboreal *cerrado* (savanna-like vegetation), suggesting a somewhat moister climate with small gallery forest, but still drier than the present one (Fig. 4.5). A paleovegetation dominated by herbaceous species is a possible explanation for the low productivity of the system represented by zone II.

Based on their similar chemical and physical properties, two different core sections have been grouped into zone II. Between the lower and the upper sections, the indicated shallow lake has undergone a process of aggradation by slope-wash sediments which are represented by the lower section of zone III (Figs. 4.2, 4.3).

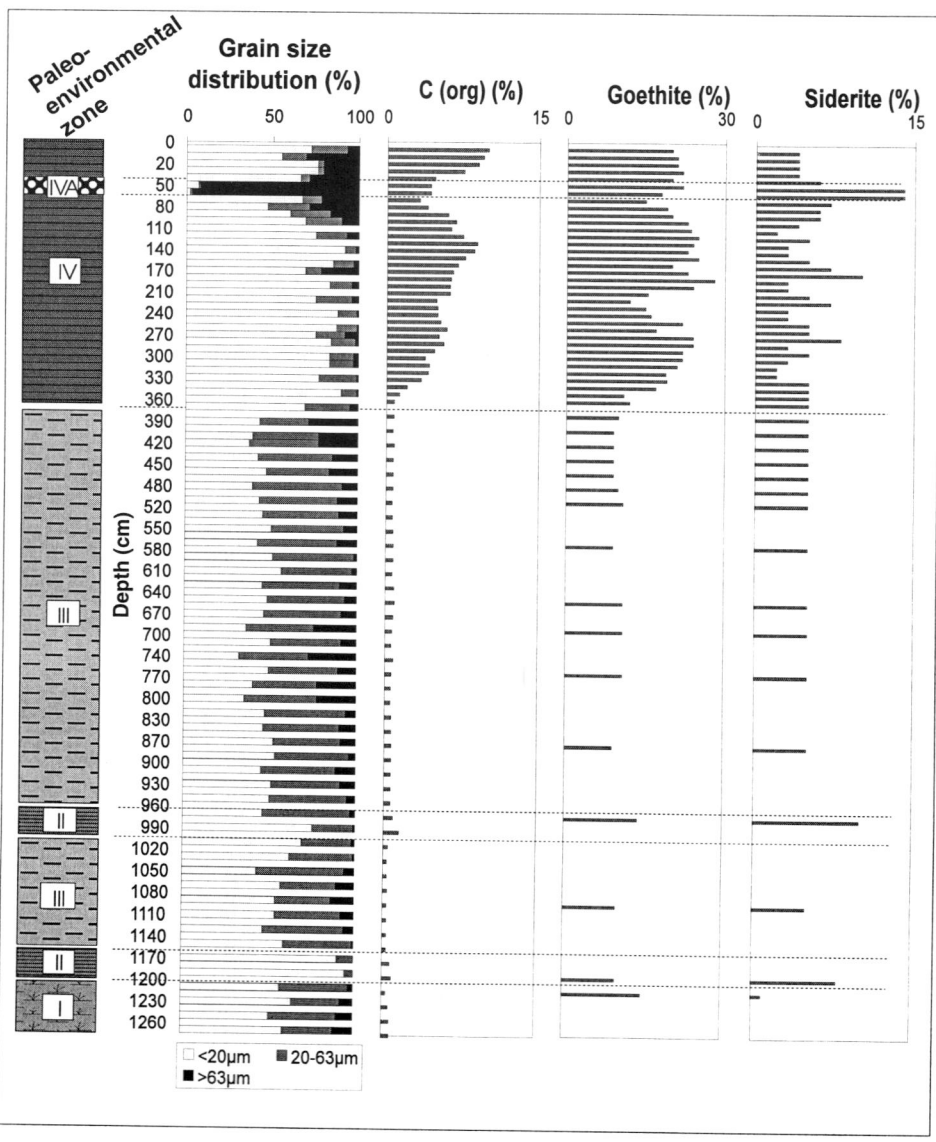

Figure 4.2 Summary diagram of mineral stratigraphy, organic carbon and grain size distribution. Paleoenvironmental zones I to IVA as described in the text.

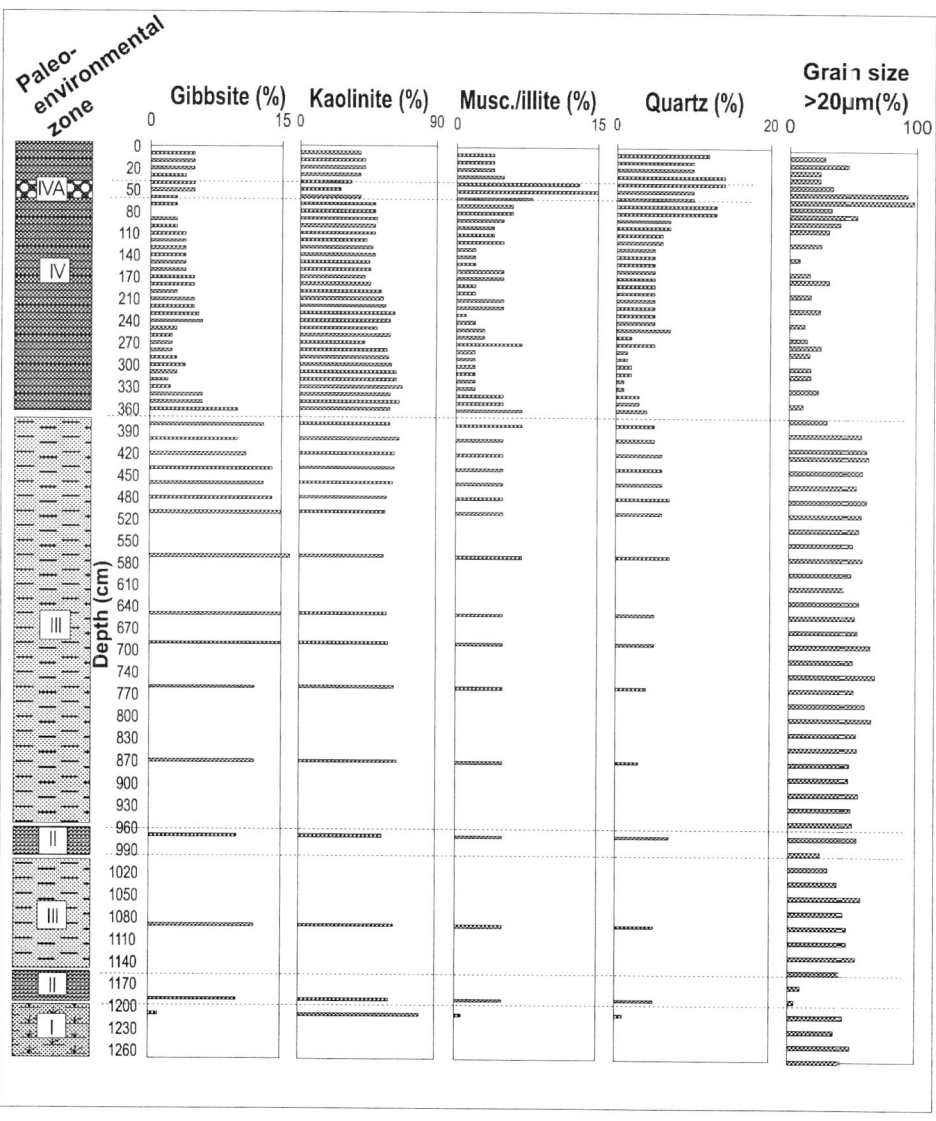

Figure 4.3 Summary diagram of mineral stratigraphy and grain size distribution.
Paleoenvironmental zones I to IVA as described in the text.

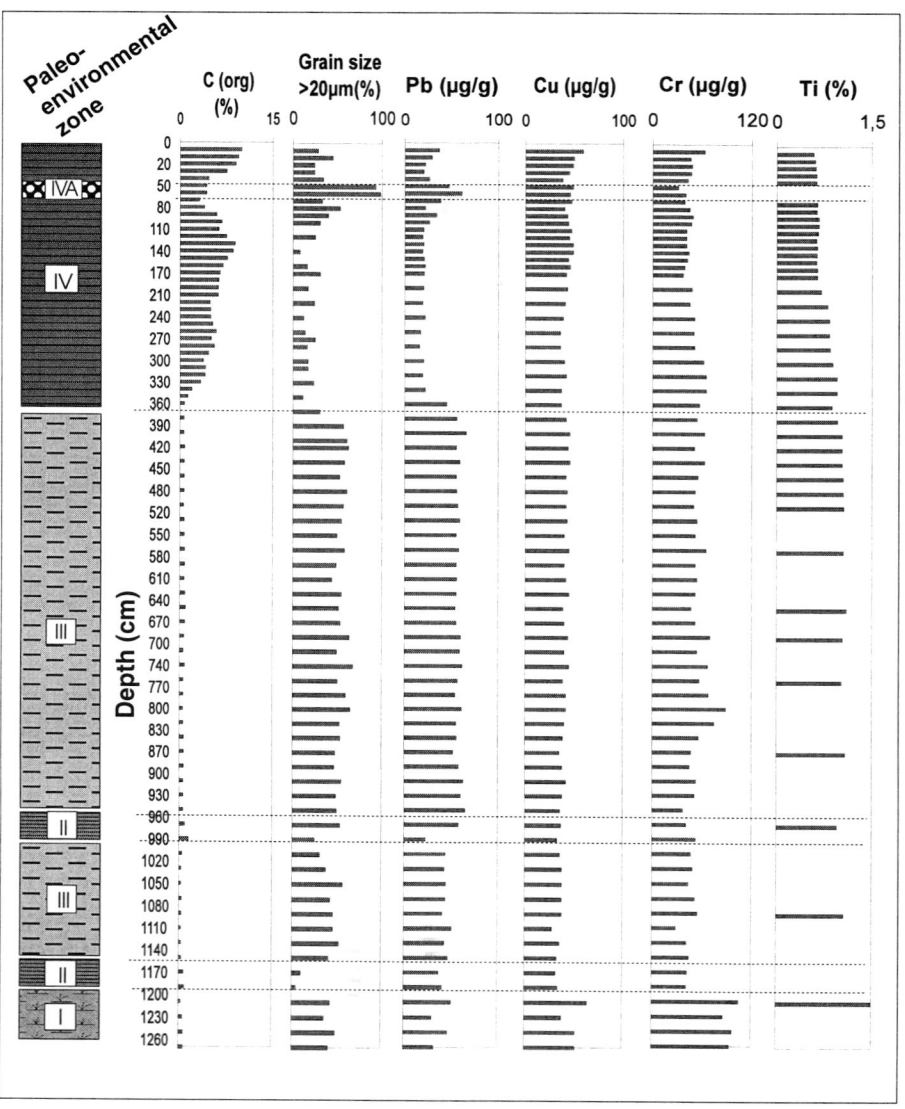

Figure 4.4 Summary diagram of chemical stratigraphy and grain size >20 μm.
Paleoenvironmental zones I to IVA as described in the text.

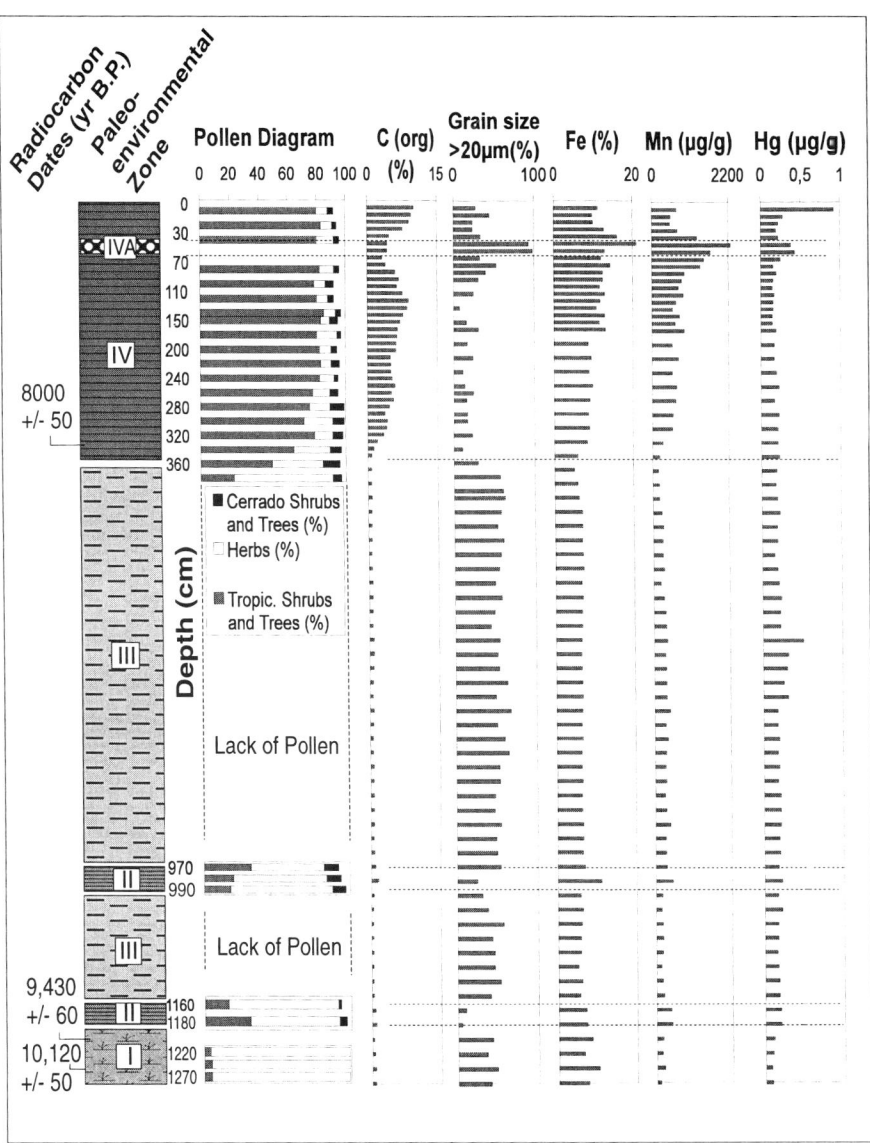

Figure 4.5 Summary diagram of chemical stratigraphy, palynological data and grain size >20 µm. Paleoenvironmental zones I to IVA as described in the text.

4.1.3 Paleoenvironmental zone III (1160-990 and 970-360 cm)

This unit comprises two core sections that together reach a thickness of 7.8 m. The sedimentary sequence consists of alternate thin layers (ca. 5 mm) of well stratified clayey and silty-sandy reddish sediments. Grain-size distribution points to the deposition of this unit during a period characterized by high transportation energy, as reflected by an increase in both the sandy fraction (average of 18%) and the 20–63 µm fraction (average of 42%). Seasonal fluctuations in the rainfall regime are likely to explain the formation of alternate clayey and silty-sandy thin layers.

A predominating allogenic nature of zone III is indicated by low contents of organic C (0.7%) and by the mineralogical composition of the <20 µm fraction, which gave as major constituents kaolinite (63%), gibbsite (15%), detrital goethite (10%), muscovite (5%) and less siderite (5%). In thin sections of the >63 µm fraction, higher muscovite contents were observed (20–40%). The deposition of muscovite-rich, silty-sandy sediments has been well traced by Pb concentrations in the <20 µm fraction, as demonstrated by their positive correlation (Fig. 4.6A). The relatively low mobility of Pb was also observed in residual soils over a base-metal deposit in North Carolina, resulting in a mainly mechanical dispersion by soil creep (Rose et al., 1979). As already observed by Kopp (1986), the behavior of Pb during weathering is closely associated with muscovite, as K tends to be replaced by Pb in the structure of muscovite due to their very similar ionic radius (Fig. 2.1). An aggrading lake basin is strongly suggested by the oxic nature of sediments and the increase in sediment coarseness, accompanied by higher gibbsite and Pb concentrations (Figs. 4.2–4.4).

A long interruption in pollen deposition was observed within the whole sequence of zone III, which is an indication of the rapid sedimentation of detrital sediments (Fig. 4.5). AMS radiocarbon dating from the base of the overlying zone IV revealed that this 7.8-m-thick sequence was indeed deposited within a very short time (Fig. 4.6D). A further indication of a major allogenic nature of this unit is the contrasting behavior of Fe relative to Mn. Fe/Al ratios present almost no variation within zone III, in contrast to Mn/Al ratios, while in the overlying organic-rich zone IV both ratios are well correlated (Fig. 4.6B, C). This indicates the occurrence of authigenic Mn and prevailing detrital Fe oxyhydroxides associated with kaolinite. Microanalyses performed by SEM with EDS confirmed the occurrence of Mn-siderite.

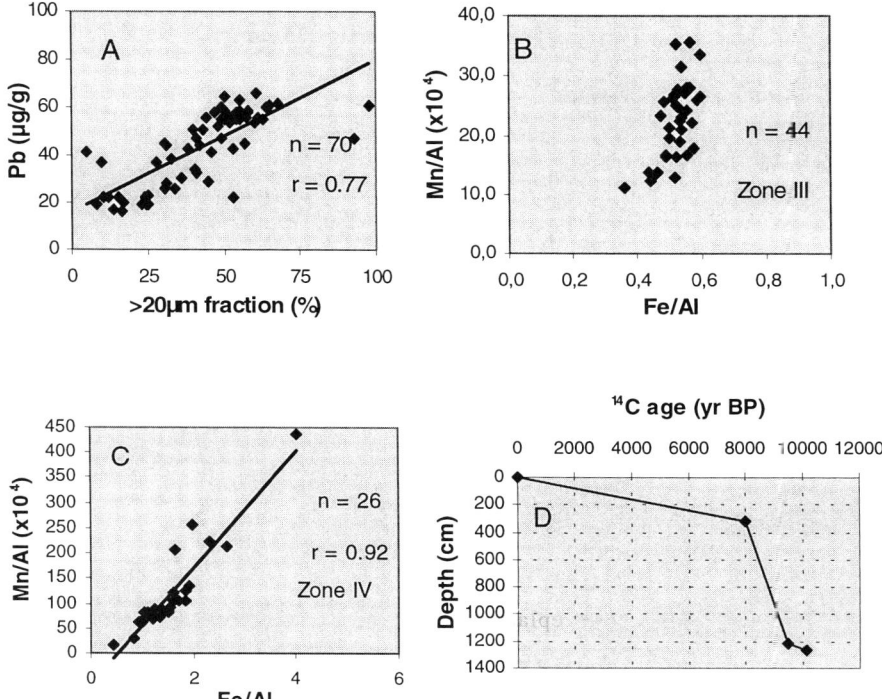

Figure 4.6 (A) Pb concentrations as a function of the >20 μm grain-size fraction; (B) relationship Fe/Al-Mn/Al ratios in zone III; (C) relationship Fe/Al-Mn/Al ratios in zone IV; (D) sedimentation rates.

4.1.4 Paleoenvironmental zone IV (360-0 cm)

A lake transgression that lasts until today is indicated by the occurrence of very fine, well-sorted sediments. This greenish sedimentary sequence is characterized by organic-rich materials structured in very thin layers (ca. 1 mm), which typify a lacustrine sedimentation. Grain-size distribution illustrates the general dominance of the <20 µm fraction (75%) accompanied by smaller amounts of both the 20–63 µm fraction (18%) and the >63 µm fraction (7%). However, contrasting patterns were observed in some layers characterized by silty-sandy sediments, which are concentrated in the upper zone IV (Fig. 4.2). Between depths of 40 and 80 cm, the occurrence of sandy materials was grouped into zone IVA.

In zone IV, sediments contain increasing organic C concentrations upwards, ranging from 1.3 to 9.9%. Mineralogical composition reflects environmental changes towards lacustrine sedimentation and more reducing conditions. This is indicated by decreasing contents of kaolinite upwards, ranging from 60 to 40%, lower gibbsite contents (average of 5%), and increasing siderite and authigenic goethite (average of 7% and 22%, respectively). The occurrence of sediments containing alternate goethite- and siderite-rich layers (Fig. 4.2) point to fluctuations in the intensity of natural eutrophic conditions that seem to be characteristic of tropical lakes (Tundisi et al., 1997).

The pollen assemblage in zone IV gave a conspicuous increase in arboreal elements (up to 85%) that represent the present-day vegetation with widespread semideciduous forest (Fig. 4.5). This vegetation change indicates a transition to semihumid climate with a dry season of 4-5 months that occurred shortly before 8000 ± 50 yr. B.P. according to AMS radiocarbon dating obtained from a well-preserved leaf (Fig. 4.7).

Figure 4.7 Photograph of the core section (340 cm) showing a well-preserved,
radiocarbon-dated leaf (core-section diameter = 4 cm).

A sharp decrease in Pb concentrations in zone IV, accompanied by slightly lower
ones of Cr and Ti, is in accordance with the increasing abundance of authigenic
materials relative to detrital sediments (Fig. 4.4). The same trend is demonstrated
by elevated Fe and Mn concentrations (Fig. 4.5). In turn, the relatively constant
values for Hg and Cu relative to the underlying unit point to the increasing
mobility of both metals, probably as a result of organically mediated redox
reactions (Figs. 4.4, 4.5). A possible indication of this association arises from the
positive correlation observed between Hg/Al ratios and Mn concentrations in zone
IV (Fig. 4.8A).

Within zone IV, three major contrasting phases were observed at 280-260, 180-
160 and 80-40 cm, where layers of silty-sandy, siderite- and muscovite-rich
sediments occur (Figs. 4.2, 4.3). The last one, named zone IVA, is composed of
sandy sediments with the predominance of quartz, siderite and muscovite. The
sharp increase in siderite contents in these layers is associated with the increasing
coarseness of sediments, as demonstrated by the correlation between siderite and

the amount of >20 μm fraction within zone IV (Fig. 4.8B). The same association with the amount of >20 μm fraction has been observed for Fe and Mn concentrations in zone IV (Figs. 4.8C, D). A negative correlation has been also observed for Fe and Mn concentrations relative to Al concentrations (Figs. 4.8E, F). This is an indication that not only siderite but also Fe and Mn oxyhydroxides are authigenic in this zone.

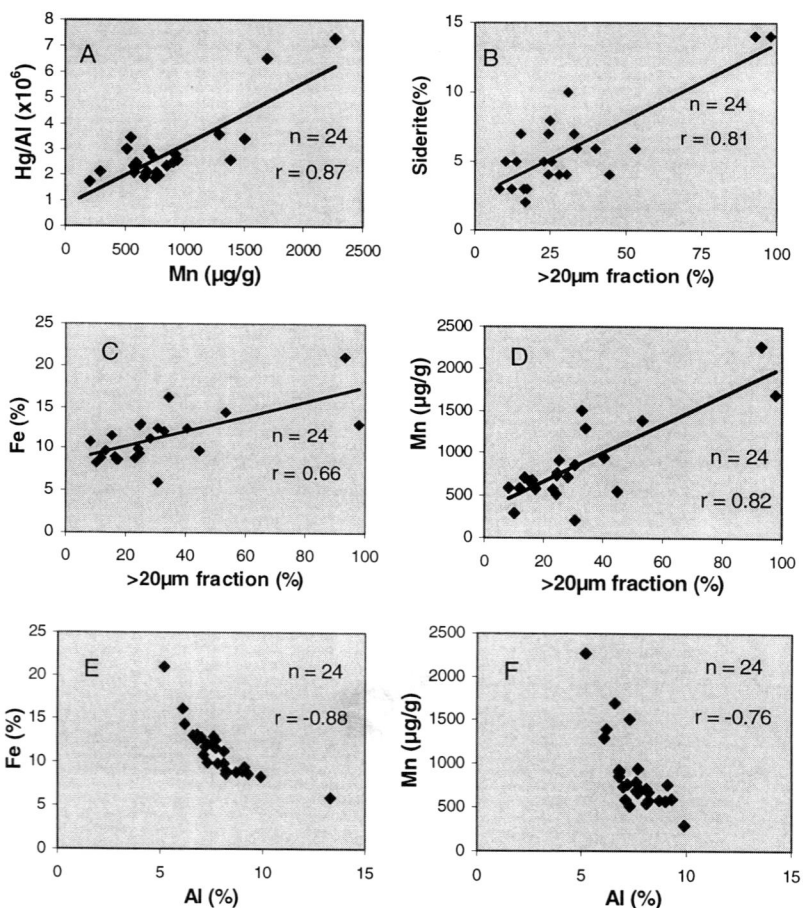

Figure 4.8 (A) Hg/Al ratios and Mn; (B) siderite and amount of >20 μm fraction; (C) Fe and amount of >20 μm; (D) Mn and amount of >20 μm; (E) Fe and Al; (F) Mn and Al.

SEM observations revealed the occurrence of siderite associated with pyrite framboids (Figs. 4.9, 4.10). The trace amounts of pyrite indicate that the rate of iron reduction fairly exceeds that of sulfate reduction, probably due to a very limited sulfate availability in the lake waters. This characteristic of the water is demonstrated by the extremely low S concentrations in sediments (see Appendix A.1). Favorable conditions for siderite precipitation from the water column may occur at the sediment-water interface during periods of lake stratification (Williamson et al., 1998; Xiouzhu et al., 1996; Moore et al., 1992; Postma, 1982; Tundisi et al., 1997). Another siderite-forming process seems to be related to early-diagenetic transformations occurring within the sediments Microanalyses using an energy dispersive unit (EDS) revealed the occurrence of Fe-rich kaolinite which appears to be a source of Fe^{2+} to the pore waters. However, the released Fe^{2+} likely reprecipitates immediately in form of siderite onto kaolinite, as demonstrated by increasing concentrations of C and Fe onto kaolinite (Figs. 4.11, 4.12). Siderite crystals commonly contain significant amounts of Mn, in the range of 5 to 10%. This point to a solid solution with $MnCO_3$, as already observed in swamp sediments of Denmark (Postma, 1982).

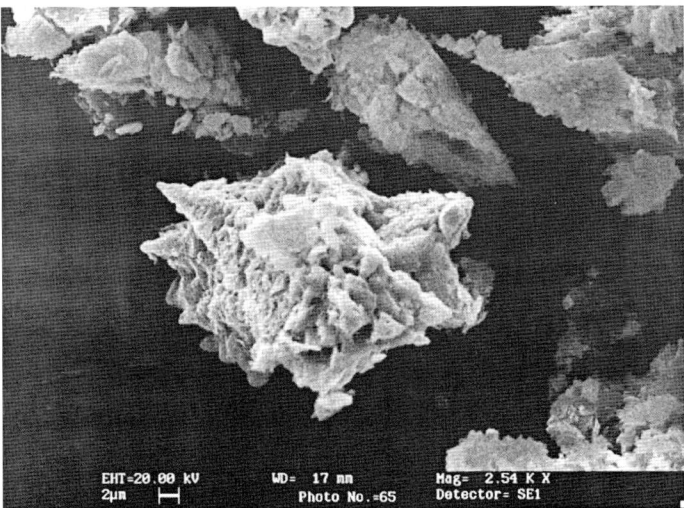

Figure 4.9 Scanning electron micrograph showing a siderite aggregate with star-shaped overgrowth.

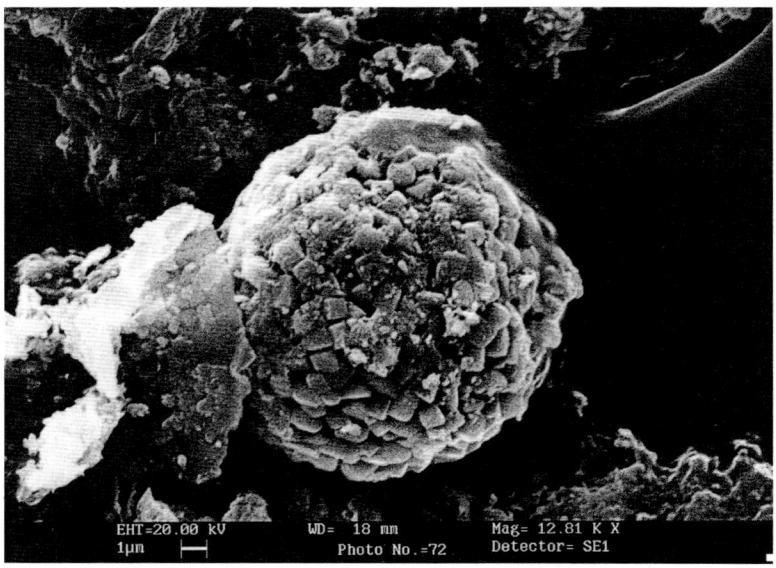

Figure 4.10 Scanning electron micrograph showing framboidal pyrite.

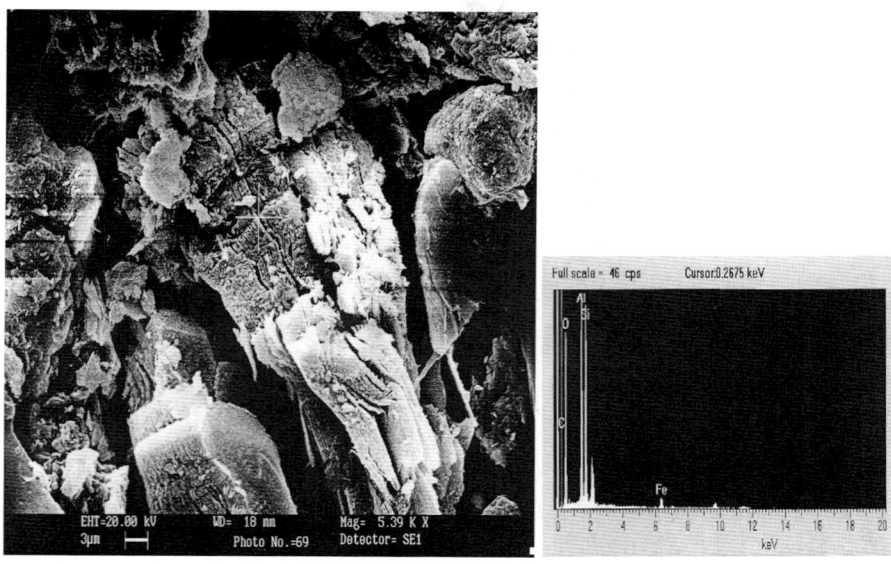

Figure 4.11 Scanning electron micrograph with EDS spectrum: Fe-rich kaolinite.

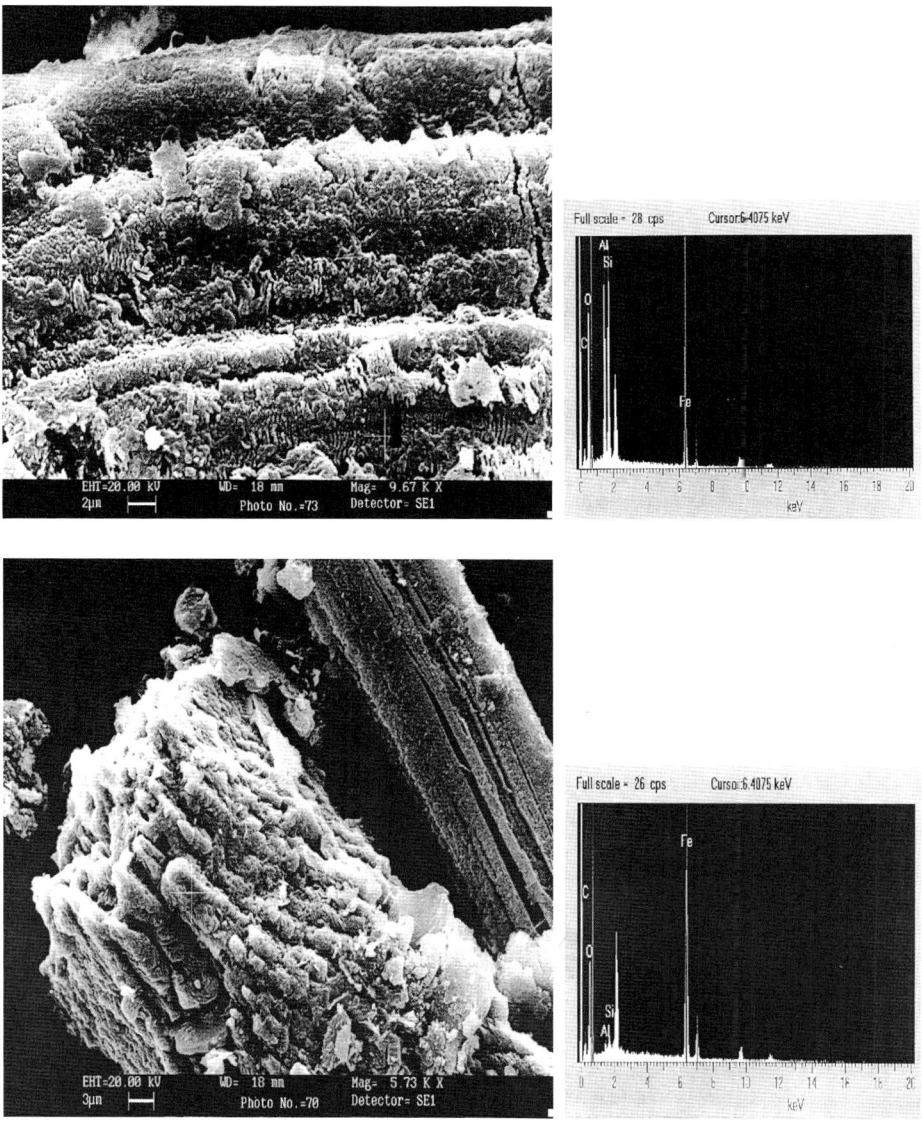

Figure 4.12 Scanning electron micrographs with EDS spectrum: Fe-rich kaolinite
showing increasing concentrations of C and Fe (lower, top; higher, bottom).

4.2 Weathering profiles EG1 and EG2

Mineralogical and geochemical contrasts within the 29-m-deep weathering profile EG2 (Fig. 4.13) allow to a certain extent to identify the source of terrigenous sediments in Lake Silvana. Profile EG1, in turn, reaches a depth of 3.5 meters and presents geochemical patterns very similar to the superficial colluvial unit of profile EG2. This indicates the widespread distribution of the colluvial unit over the river basin.

The lowermost C2 horizon of profile EG2 consists of brownish-ochre saprolite without banding, indicating a somewhat isotropic fabric of the parent rock. Saprolite is possibly derived from the decomposition of an amphibolite lens, as indicated by very low quartz contents. Samples are composed of poorly crystallized kaolinite, goethite, hematite and quartz. Moreover, samples from the C2 horizon give the highest values for Ni, Cu, Zn, Ti and Fe (Fig. 4.14). High Fe contents are probably derived from the decomposition of mafic minerals found in the parent rock. Petrographically, the occurrence of Fe oxyhydroxides and poorly ordered kaolinite as pseudomorphs after biotite could be observed.

The ordering degree of kaolinite was determined through X-ray diffractograms, which clearly show two different patterns for poorly and well ordered kaolinite (Fig. 4.15). The strong displacement of the b-axis in poorly ordered kaolinite appears related to the water activity, as its occurrence is restricted to the surface horizon (colluvial unit) and to the C2 horizon, located near the groundwater table. In the B and C1 horizons, where a lower water activity is likely, well ordered kaolinite occurs.

Figure 4.13 Weathering profile EG2 showing the colluvial unit at the top.

The C1 horizon consists of reddish-brown saprolite showing the original macroscopic structure of the parent rock, a well defined banding characteristic of the biotite-rich gneiss. Samples are composed almost exclusively of well ordered kaolinite, hematite, quartz and to a lesser degree, goethite. The geochemical and mineralogical composition, with more quartz and less Fe and Ti relative to the underlying unit, supports the assumption of a saprolite derived from the biotite-rich gneiss. This interpretation is supported by another study on the behavior of Fe and Ti in saprolites relative to the parent rock. Kopp (1986) has observed that Fe and Ti concentrations in saprolites of weathering profiles in Minas Gerais show almost no variation relative to the parent rock. Trace-metal concentrations in the C1 horizon revealed the highest values for Cr and Pb (Fig. 4.14). As already observed elsewhere, the first stage of chemical weathering in tropical landscapes results in the formation of kaolinite and amorphous or crystallized ferruginous oxyhydroxides, preserving the macroscopic lithological structure (Nahon, 1991).

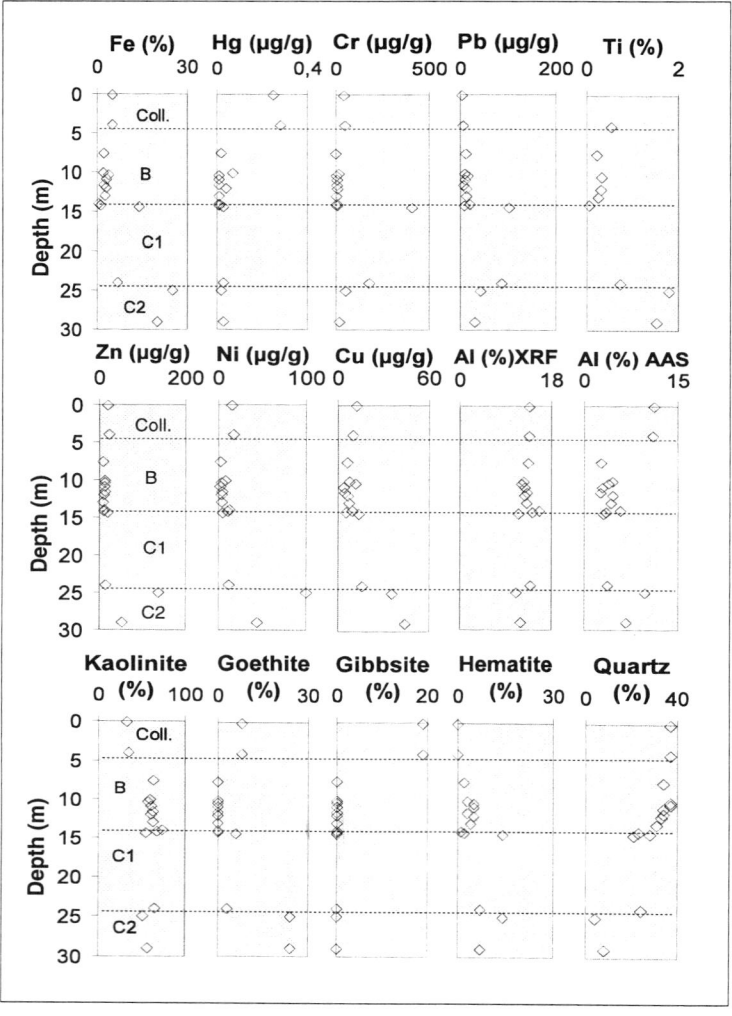

Figure 4.14 Metal and mineral distribution within the weathering profile EG2.

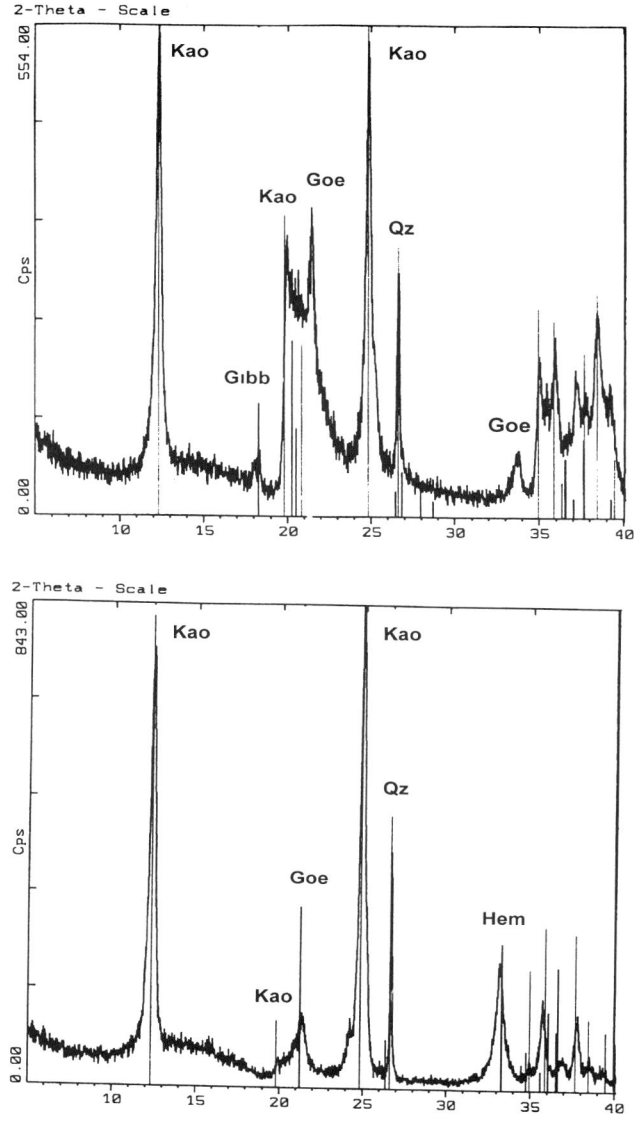

Figure 4.15 Typical diffractograms of poorly (top) and well ordered kaolinite (bottom)-Weathering profile EG2.

The B horizon consists of bleached-reddish soils composed by well-ordered kaolinite, quartz and hematite. Above the groundwater table, the decreasing water activity seems to have caused the complete crystallization of kaolinite and consequently its deferruginization. This Fe-depleted horizon also corresponds to the segment with the lowest metal concentrations, indicating their association with Fe oxyhydroxides, which were probably removed into solution as Fe^{2+}. Tardy and Roquin (1992) have summarized the main factors which control the occurrence of goethite, hematite and kaolinite in lateritic profiles. Soils have generally higher hematite/goethite ratios with increasing dryness, as observed in the B and C1 horizons, an undersaturated zone from depths of 14 to 24 meters. Furthermore, Tardy and Roquin (1992) suggested that the iron content in kaolinite is dependent on the water and dissolved silica activity. An increase in hydrolysis and/or silica activity should induce an increase in iron relative to aluminum contents in ferruginous kaolinite.

From the surface down to 4 meters, a colluvial unit (Coll.) covers the lateritic sequence. It consists of a highly leached, yellowish sandy layer composed mainly of poorly ordered kaolinite, corroded quartz grains, gibbsite and goethite (Fig. 4.14). It is interesting to note the relatively high Hg concentrations and gibbsite contents associated with very low Pb and Cr concentrations. Very similar features have been observed in the Amazonian lowlands, where ochre-coloured clays with different amounts of sand occur in the uppermost horizon above Precambrian shields and Palaeozoic sediments (Irion, 1984). The ochre horizon is widely distributed over the Amazonian lowlands and contains poorly ordered kaolinite, in contrast to lower horizons, corroded quartz, gibbsite and goethite and exhibits a relative enrichment of Cr and Ti. During the wet season the quasi-saturation of the surface horizons acts in favor of the rehydration of minerals previously dehydrated during the dry season. This leads to the formation of goethite and gibbsite (Tardy and Roquin, 1992). At a depth of 4 meters, an erosional discontinuity reveals the allochthonous character of this unit, which seems to have been deposited after downslope colluvial transport of the former upper horizon of the lateritic profile. As mentioned already, the profile EG1 has geochemical patterns that resemble those of the colluvial unit (Figs. 4.14, 4.16).

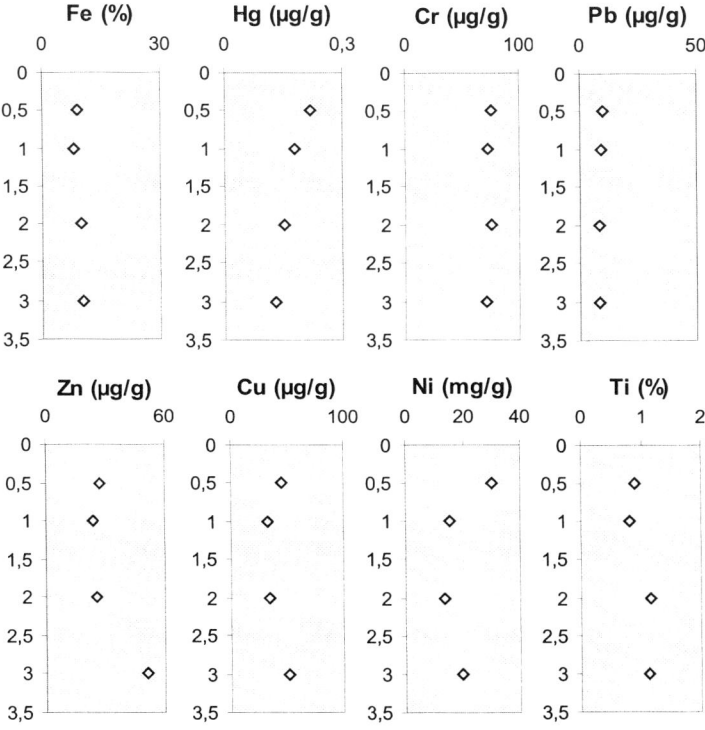

Figure 4.16 Metal distribution within the weathering profile EG1.

Although Al concentrations measured by XRF do not vary significantly within the profile EG2, those obtained by AAS after acid digestion with *aqua regia* do. This extraction procedure appears to be pseudo-selective for the extraction of Al incorporated in gibbsite and poorly ordered kaolinite. This has been indicated by comparing the extraction rates of Al from both weathering profiles and lake sediments, by means of the *aqua-regia*-soluble Al/Al (XRF) ratio. In sediments containing poorly ordered kaolinite and gibbsite this ratio is almost constant with a mean value of 0.70 (Fig. 4.17). Within the weathering profiles the extraction rates range from 0.20 to 0.37 in the B and C1 horizons, with well ordered kaolinite and no gibbsite, and from 0.56 to 0.88 in the colluvial unit and C2 horizon (Table 4.1). Thus the extraction rates of Al by *aqua regia* show a positive correlation with contents of both poorly ordered kaolinite and gibbsite.

Table 4.1 Metal concentrations within the weathering profiles EG2 and EG1

Profile/ Horizon	Depth (m)	Hg	Pb	Zn	Ni	Cu	Cr	Mn	Fe %	Al XRF %	Al AAS %	Al AAS/ XRF	Ti %
		<---- ----- ------- µg/g ------- ------ ---->											
EG2/Coll.	0.1	0.25	3.5	19.5	15.5	12.5	42	22	3.8	--	11.2	--	--
EG2/Coll.	4.0	0.28	5	22	17	10	48	27	3.8	13.6	11.1	0.80	0.54
EG2/B	7.6	0.02	12.5	8	2	6	1.5	12	2.0	13.4	2.7	0.20	0.22
EG2/B	10.1	0.07	9	12.5	8	7.5	16	23	3.1	--	4.5	--	--
EG2/B	10.4	0.01	15	12	4	11	2	31	2.7	12.1	3.9	0.32	0.33
EG2/B	10.9	0.01	11	11.5	2.5	3.5	5.5	14.5	2.2	--	2.9	--	--
EG2/B	11.6	0.01	7	13	4	4	5	25	0.8	--	2.6	--	--
EG2/B	12.0	0.04	13	11	3	6	8	30	2.9	12.7	4.6	0.36	0.31
EG2/B	13.0	0.01	13	8	4	7	3	6	1.9	13.1	4.3	0.33	0.25
EG2/B	14.0	0.01	20	11	12	9	6	5	0.4	15.6	5.8	0.37	0.05
EG2/B	14.2	0.01	8	11	10	5	5	1	1.3	--	3.5	--	--
EG2/C1	14.4	0.03	102	19	5	14	409	35	11.6	--	3.1	--	--
EG2/C1	24.0	0.03	87	14	12	16	178	49	6.9	13.8	3.7	0.27	0.73
EG2/C2	25.0	0.02	42	137	100	35.5	53	67.5	21.8	11.0	9.7	0.88	1.79
EG2/C2	29.0	0.03	30	52	44	44	18	62	15.4	11.9	6.7	0.56	1.52
EG1/Coll.	0.5	0.22	10	28	30	45	75	143	8.9	--	12.3	--	--
EG1/Coll.	1.0	0.18	9.5	23.5	15.5	32	73.5	113	8.1	--	11.9	--	--
EG1/Coll.	2.0	0.15	8	26	14	35	75	133	9.8	--	11.1	--	--
EG1/Coll.	3.0	0.13	9	52	20	51	70	285	10.3	--	11.4	--	--

-- not measured

A light increase in Fe, Ti and Cr concentrations were also observed in the colluvial unit relative to the B horizon. It is well known that the advanced weathering stage is normally accompanied by the relative accumulation of less mobile metals under oxidizing conditions and neutral pH such as Al, Fe, Ti and Cr (Rose et al., 1979). Both the weathering degree and the composition of the parent rock are controlling the metal distribution within the weathering profile EG2, with the possible exception of Hg. The increase in gibbsite contents at the surface, in this case accompanied by elevated Hg concentrations, is a typical picture of highly leached soils. High concentrations of Ti, Cr, Pb, Ni, Cu and Zn in the C1 and C2 horizons seem mainly influenced by the mineralogical composition of the parent rock.

The high average concentration of Hg in the colluvial unit (0.20 µg/g) represents an increase to values 10 times higher than those observed in the saprolite. Hg in the colluvial unit is likely to be incorporated in gibbsite, since a positive correlation (r = 0.83) exists between Hg and *aqua-regia*-soluble Al in both weathering profiles (Fig. 4.17), while almost no correlation between Hg and Fe exists (Fig. 4.14).

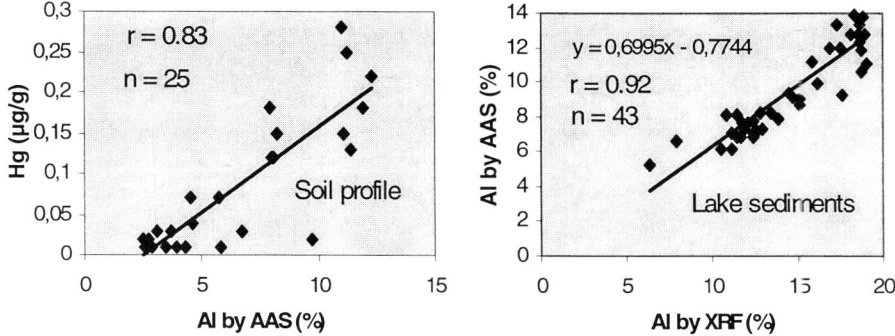

Figure 4.17 Correlation between Hg and *aqua-regia*-soluble Al in both weathering profiles and between *aqua-regia*-soluble Al and Al (XRF) in lake sediments.

4.2.1 Mercury accumulation at the top of profiles

A study assessing mercury pollution in two gold mining regions of the Brazilian Amazon has shown that Fe-rich soils and sediments play a major role in retaining/transporting Hg (Rodrigues-Filho and Maddock, 1997). There, a possible association between Hg and Al hydroxide was not taken into consideration. Roulet et al. (1996) have observed an important accumulation of Hg in surface layers of different Amazonian soils, reaching the same levels as in the present study (0.20 μg/g). The authors pointed out that these Hg levels are an order of magnitude greater than those reported for temperate and nordic soils. Furthermore, the Hg accumulation appeared to be entirely controlled by Fe and Al oxyhydroxides. The closely associated contents of Fe and Al oxyhydroxides in all samples did not allow the authors to evaluate whether Hg is enriched on Fe or Al oxyhydroxides, or both. The present results, in turn, suggest that the widespread Hg accumulation in surface horizons of Brazilian lateritic soils may be mainly controlled by Hg adsorption onto gibbsite. However, to test this hypothesis more detailed studies are required.

It is noteworthy that a presumable Hg immobility during weathering, as indicated by its accumulation in surface horizons, is not in accordance with its ionic potential which points to its high mobility (Fig. 2.1). Indeed, metallic Hg (Hg^0) seems to be easily oxidized to inorganic Hg salts, notably halides and sulfates, as a result of natural leaching or weathering cycles (Jonasson and Boyle, 1979). Based on data from Canadian soils, these authors have demonstrated that Hg concentrations tend to be enriched in surficial humic soils presumably due to litter decomposition, but with enrichment factors relative to the saprolite horizon not greater than 2, while Hg concentrations in the B horizon exhibit no enrichment relative to the saprolite horizon. Relatively high background concentrations of Hg (average of 0.15 μg/g) were recorded in B horizons of undisturbed soils in Central Brazil, which have been demonstrated to be derived from the oxidation of Hg-containing pyrite (Rodrigues-Filho and Maddock, 1997). There, Hg concentrations tend to decrease from the saprolite to the top of profiles, demonstrating that Hg concentrations have been depleted through leaching (Rodrigues-Filho, 1995).

Another source of Hg in soils is the atmospheric Hg which may be naturally released from evapo-transpitation of leaves, decaying vegetation and volcanic activity (Jonasson and Boyle, 1979). Hg is also released to the atmosphere through anthropogenic emissions, among which gold mining and deforestation seem to be the most relevant in the Brazilian Amazon. Estimates of the Hg emissions derived from gold mining in this region point to 200,000 to 260,000 tonnes Hg since the beginning of the European colonization (Malm, 1998; Lacerda, 1997). The amount of Hg emitted by deforestation has been estimated from the biomass distribution in the Amazon, reaching a value of 710 tonnes Hg for the last 20 years (Veiga et al., 1994). As tropical ferralitic soils are thought to be as old as ~ 5 Ma. (Nahon, 1986; McFarlane, 1983), the long-term deposition of atmospheric Hg in soils rich in Fe and Al hydroxides is likely to explain the widespread Hg accumulation in surficial ferralitic soils rather than a weathering-driven accumulation process from the parent rock. If so, most of the Hg found in surficial soils is derived from natural rather than anthropogenic sources, as sediments with elevated Hg concentrations have been deposited in Lake Silvana since, at least, 9000 yr ago (Fig. 4.5).

4.3 Physical and chemical parameters of the water column of Lake Silvana

Physical and chemical parameters of the water column above the core site were obtained by *in situ* measurements of Eh, pH, electrical conductivity, dissolved oxygen and temperature in summer time (Dec. 1996). During winter (June 1996), dissolved oxygen and electrical conductivity were not measured. In addition, laboratory analyses of dissolved Fe and Mn were performed in samples collected during the summer.

No accentuated thermal stratification was observed within the water column in summer, with a decrease of only 1°C from the epilimnion to the hypolimnion. There is a clear chemical stratification, however, characterized by the progressive oxygen consumption from the surface. In winter (June 1999), temperature was constant and relatively low throughout the water column, while pH and Eh indicate neutral and oxidizing conditions (Fig. 4.18).

Thermal stratification in the Rio Doce lakes is commonly characterized by a decrease in temperature of 5 to 8°C during the summer, while most lakes present no stratification during the winter (Tundisi et al., 1997). Based on limnological data from 15 lakes of the Rio Doce lake system, these authors suggested that lakes with a maximum depth of 10 meters, like Lake Silvana, tend to present no thermal stratification in the summer due to continuous circulation processes. The effectiveness of this circulation is a function of wind force, surface cooling and lake depth.

An accentuated oxygen consumption beginning at a depth of 2 m demonstrates a particular characteristic of Lake Silvana relative to other lakes of this region, where O_2 consumption commonly starts at 4 m or deeper (Tundisi et al., 1997). This is an indication that either respiration rates or decay of terrestrial organic matter, or both, in Lake Silvana exceed those from all other lakes. The oxycline in the 10-m deep Lake Carioca most closely resembles that of Lake Silvana. Stratification in Hall Lake, a soft-water lake showing permanent stratification and located to the north of Seattle, USA, also presents an oxycline and pH range quite similar to Lake Silvana (Balistrieri et al., 1994).

In summer, pH values in Lake Silvana typically decrease across the zone of aerobic oxidation of organic matter and undergo a slight decrease in the anoxic zone. Values of pH ranged from 7.8 at the surface to 7.2 at the bottom of the water column, while electrical conductivity varied from 79 to 155 μS/cm near the sediment-water interface (Fig. 4.18). The other lakes in this region show values of pH and electrical conductivity significantly lower than in Lake Silvana; pH values commonly range from 7.0-6.5 at the surface to 6.3-6.0 at the bottom, while electrical conductivity ranges from 20-30 to 50-70 μS/cm downwards. Electrical conductivity, alkalinity and concentrations of NH_4^+ and ΣCO_2 are very well correlated in the water column of Lake Carioca and other lakes studied by Tundisi et al. (1997). Drever (1982) pointed out that at constant CO_2 activity and in the absence of anions such as Cl^-, SO_4^{2-} and NO_3^-, the total cation concentration will be approximately equal the bicarbonate concentration, and hence pH is inversely related to salinity and alkalinity. For the system H_2O-CO_2, the author presents the following charge balance equation:

$$m (H^+) = m (HCO_3^-) + 2m (CO_3^{2-}) + m (OH^-)$$

where m is the concentration. If we introduce sodium into the system, the charge balance equation becomes:

$$m (H^+) + m (Na^+) = m (HCO_3^-) + 2m (CO_3^{2-}) + m (OH^-)$$

For solutions that are relatively neutral, $m (H^+)$, $m (OH^-)$ and $m (CO_3^{2-})$ are generally negligible compared to $m (Na^+)$ and $m (HCO_3^-)$, which will be equal. Thus, pH increases with salinity. This is a possible explanation for the higher pH values in Lake Silvana, accompanied by higher electrical conductivity, in comparison to the above-mentioned lakes. However, within the water column of Lake Silvana, the decreasing pH values with depth are probably the net effect of the acid production from oxidation of organic matter and the increasing electrical conductivity, yielded by the base-producing reduction of FeOOH and NO_3^-.

Relatively high concentrations of dissolved Fe^{2+} and Mn^{2+} – 0.2×10^{-3} M and 0.15×10^{-4}, respectively – have been measured in the bottom waters of Lake Silvana in the summer, corresponding to the zone of highest ammonium concentration (0.11×10^{-3} M) in Lake Carioca. At a depth of 5 meters, dissolved concentrations decreased to 0.2×10^{-4} M for Fe^{2+} and 0.15×10^{-5} M for Mn^{2+}, whereas at the surface concetrations were not detectable. Likewise, Balistrieri et al. (1994) demonstrated that a sharp increase in concentrations of dissolved Fe^{2+} (up to 0.16×10^{-3} M) occurs parallel to the curve of ammonium production, below the zone of sulfate reduction in Hall Lake, which in turn, contrasts with the predictable onset of nitrate reduction prior to sulfate reduction. The occurrence of high Fe^{2+} concentrations in the zone of ammonium production is in accordance with the thermodynamically-calculated similar efficiency in decomposing organic matter for both the reduction of FeOOH and NH_4^+ production. (Stumm and Morgan, 1996).

5 Discussion

Holocene climatic changes in southeastern Brazil were investigated through the correlation of paleoenvironmental proxies (geochemistry, mineralogy and pollen) from the sediment core SB1. Contrasting geochemical and mineralogical patterns have been attributed to changes in both lake level and slope-erosion activity; the first causing variations in the concentration of authigenic Mn, goethite and siderite, and the latter leading to contrasting inputs of silty-sandy sediments and detrital Cr, Pb, Hg, gibbsite and muscovite.

For the phase preceding the natural damming that according to Pflug (1969) gave rise to the Rio Doce lakes, sediment composition indicates a source from lower horizons of the weathering mantle, while during phases of intense slope-erosion activity, detrital sediments are likely derived from the mixing of horizons regardless of their topographic position. In turn, sediments deposited during phases of less intense erosion and lacustrine sedimentation present no evidence of source, reflecting their major authigenic nature.

Sediments in zone I present trace-metal concentrations that resemble those from the C1 and C2 horizons in the lower weathering profile, with elevated Cr, Pb, Cu and Ti, and very low Hg concentrations (Table 5.1). Mineral composition also indicates this source, as ca. 95% of the mineral contents in this zone are characterized by kaolinite and goethite, while gibbsite and quartz are absent (Fig. 4.3, Table 5.1). The low contents of organic matter, concentrated as plant rootlets, and the absence of siderite together point to an oxic environment (Fig. 4.2).

With the absence of characteristic imprints from the uppermost horizon of the weathering mantle (colluvial unit), namely Hg and gibbsite, the pollen-indicated sparse vegetation and dry climate strongly suggest that the drainage basin was still open when this sequence formed. Otherwise materials derived from less protected superficial soils would inevitably have accumulated in the basin. These features likely represent a buried paleosol or a floodplain deposit preceding the damming of the drainage basin.

Table 5.1 Comparison of mean concentrations of selected trace metals and secondary minerals within zone I with the weathering mantle

Sample type	Cr (µg/g) mean (σ)	Pb (µg/g) mean (σ)	Cu (µg/g) mean (σ)	Hg (µg/g) mean (σ)	Gibbsite (%) mean (σ)	Kaolinite (%) mean (σ)
Zone I * (core SB1)	96 (±6.8)	40 (±9.1)	52 (±9.2)	0.09 (±0.01)	n.d.	80 (--)
Zone II * (core SB1)	45 (±6.1)	34 (±11.2)	34 (±1.5)	0.20 (±0.02)	10 (±0.0)	57 (±2.5)
Zone III * (core SB1)	54 (±10.2)	55 (±3.4)	41 (±2.8)	0.21 (±0.06)	14 (±1.7)	61 (±3.1)
Colluvial unit (EG1+EG2)	64 (±12.5)	7.5 (±3.9)	31 (±15.4)	0.20 (±0.04)	19 (±0.0)	34 (±1.0)
B horizon (profile EG2)	5.6 (±4.1)	12 (±3.6)	6.8 (±2.3)	0.02 (±0.02)	n.d.	64 (±4.5)
C1 + C2 horiz. (profile EG2)	165 (±153)	65 (±30)	27 (±12.8)	0.03 (±0.01)	n.d.	57 (±4.8)

σ = standard deviation; n.d.= not detectable; * sediment fraction <20 µm

In zone II, lower Cr, Pb and Cu concentrations and higher gibbsite and Hg concentrations appear to be mainly derived from the colluvial unit covering the pedologic mantle (Table 5.1, Figs. 4.3–4.5). This supports the interpretation of lake formation, as sediments likely derived from the uppermost unit of the pedologic mantle started to accumulate in the drainage basin. The indicated formation of a shallow lake shortly after 9430 yr B.P. is possibly related to the natural damming of tributary valleys through alluvial sediments, as described by Pflug (1969). However, geochemistry and pollen data point to a very early stage of the climate-driven geomorphic process that culminated in the formation of the present lake system.

In zone III, Pb and Hg concentrations point to more than one source, namely from the C1 horizon and the colluvial unit (Table 5.1). Alternate thin layers of silty-sandy sediments, with elevated Pb concentrations, and clayey sediments containing Hg, gibbsite and siderite probably reflect seasonal fluctuations in rainfall magnitude, leading to alternate erosion patterns: less intense erosion related to the deposition of fine sediments derived from the superficial colluvial unit, and strengthened debris flow of coarser material derived from the C1 horizon (Figs. 4.2–4.4). As already described elsewhere, this is a typical

Holocene sequence of slope-wash deposition interfingered with lacustrine sediments in the middle Rio Doce lowlands (Meis and Monteiro, 1979).

This seems to confirm a period of major denudational activity on the hillslopes, whose age can be estimated at 8500 yr. B.P. (Fig. 5.1). The deposition of both slope-wash sequences of zone III was interpreted as a consequence of two extreme episodes of increasing rainfall during a climate transition towards moister conditions. Slope erosion on the catchment of Lake Silvana appears to have been favored by the sparse vegetation indicated for zone II (tropical savanna).

The present interpretation is supported by two apparently contradictory assumptions of changes in the grain-size distribution of clastic sediments as a response to paleoclimate fluctuations. On one hand, the deposition of coarse clastics has been demonstrated to reflect dry periods (Ledru et al., 1998; Turcq et al., 1997; Martin et al., 1993). This association has generally been explained with the increasing proximity to shoreline, the sparse vegetation and low lake levels, favoring the deposition of coarser detrital sediments. On the other hand, as inflow discharge and rainfall magnitude increase, coarser sediments can be transported into small lakes, giving rise to a sedimentological pattern associated with moister conditions (Campbell, 1998; Eden and Page, 1998). These studies have demonstrated the validity of this assumption by comparing the grain-size record with known historic climate fluctuations in Canada and New Zealand, respectively. The present results suggest that the onset of the deposition of silty-sandy, slope-wash sediments in Lake Silvana is likely a result of an abrupt transition from dry to moister conditions rather than a steady-climate period, thus allowing the combination of both assumptions. The short duration of this climatic event, possibly less than 500 years as indicated by AMS radiocarbon dating, constitutes an additional indication of a transitional process (Fig. 5.1).

5.1 Correlation with global-scale climatic events

A shift from dry to moister climates has been inferred from sedimentological and palynological records in southern (Ledru et al., 1994) and southeastern Brazil (Behling, 1995; Servant et al., 1989) and in southeastern Africa (Williamson et al., 1998) at 8000, 8800, 8500 and 7800 yr B.P., respectively. Behling (1995) pointed out that a major increase in moisture, as indicated by a high resolution palynological record, lasted from 8800 to 7500 yr B.P. in Lago do Pires (17°57'S, 42°13'W), southeastern Brazil. In Central Brazil (17°15'S, 49°20'W), however, a geochemical and palynological record indicates that the Holocene transition towards moister climates occurred at 6500 yr B.P. (Salgado-Labouriau et al., 1997).

All these observations point to a large-scale increase in effective moisture at southern low to mid-latitudes, which is possibly related to the early to mid-Holocene climatic event registered in ice cores from Greenland (Stager and Mayewski, 1997; Alley et al., 1997; O'Brien et al., 1995; Blunier et al., 1995; Street-Perrott, 1993). These authors have identified an abrupt early to middle-Holocene climatic fluctuation based on different records from ice cores, including records of atmospheric methane, oxygen isotope and marine aerosols, such as calcium and sodium. It has been demonstrated that greater atmospheric aerosol loadings correspond to drier periods in the Northern Hemisphere (O'Brien et al., 1995), whereas lower values in the record of atmospheric methane are believed to represent periods of increasing dryness at low latitudes (Street-Perrott, 1993).

The records from ice cores, however, exhibit chronologically juxtaposed peaks in different paleoenvironmental proxies that strongly indicate a period of drought, probably widespread over the Northern Hemisphere, thus corresponding to an inverse fluctuation relative to that indicated for the study area. Records of atmospheric methane are likely less influenced by changes in the hydrological cycle at southern low latitudes, as the larger northern continents suggest. Through a paleoclimatic model , a large-scale decrease in effective moisture at northern low latitudes has been explained with an orbitally induced decrease in summer insolation (July to August) between 9000 and 6000 yr B.P., which would have the opposite effect in the southern tropics (COHMAP, 1988). Although the present results corroborate the model's simulation of the nature of a major Holocene climatic fluctuation for the Southern Hemisphere, it appears to have occurred shortly before 8000 yr B.P., as supported by more recent observations (Stager and Mayewski, 1997; Alley et al., 1997; O'Brien et al., 1995; Blunier et al., 1995; Street-Perrott, 1993). The present interpretation of a major climatic shift in Lake Silvana as a response to an orbitally induced increase in summer insolation is strongly supported by paleoenvironmental proxies from Lake Tritrivakely, Madagascar, where the advent of moister conditions seem to have occurred from ~7800 yr B.P. (Williamson et al., 1998).

Pflug (1969) has investigated the possible origin of the lake system through the fluviatile deposits on the Rio Doce basin, whose stratigraphy suggests that the river channel has undergone a process of aggradation by coarse sediments. This process is thought to have occurred during a semi-arid period in the late Pleistocene. A single radiocarbon dating (14,160 ± 500 yr B.P.) was obtained from carbonaceous matter occurring in the coarse clastics of the alluvial terrace, therefore the possibility of sediment reworking should be considered. Bigarella and Andrade-Lima (1982) attribute the occurrence of the Brazilian Holocene terraces to climatic fluctuations with forest retreat favoring slope erosion.

Although a climate-driven process of aggradation by alluvial/colluvial deposits does appear to have led to the clogging of the basin's outlet, our results point to a process starting around 9000 yr B.P. with the formation of a shallow lake, and culminating around 8500 yr B.P. with the flooding towards the present level (Figs. 5.1, 5.2). Our results suggest that the silty-clayey Holocene terraces, which are widespread in the southeastern Brazilian landscape (Bigarella and Andrade-Lima, 1982; Modenesi, 1988), formed during extreme rainfall episodes in the early to mid-Holocene climatic transition from dry to moister conditions. Similarly, another sedimentological study from a site ca. 300 km further to the north attributes the occurrence of alluvial fans damming tributary valleys to an early-Holocene dry period followed by concentrated rainfall, according to three radiocarbon dates between 9500 and 8700 yr B.P. (Servant et al., 1989).

Since the beginning of zone IV, a decrease in hillslope denudation between 8500 and 8000 yr B.P. is expressed by the drastic decrease in the sedimentation rate, which appears associated with the expansion of the vegetation cover (Fig. 4.5). The age of forest expansion around Lake Silvana resembles that observed in other palynological records from southeastern (Behling, 1995) and southern Brazil (Ledru et al., 1994), where forest expanded around 8800 and 8000 yr B.P., respectively. Similarly, a sedimentological study in another lake from the Rio Doce basin revealed a radiocarbon date of 7840 ± 250 yr. B.P. for the base of the uppermost 3-m-thick peat sequence (Meis and Monteiro, 1979).

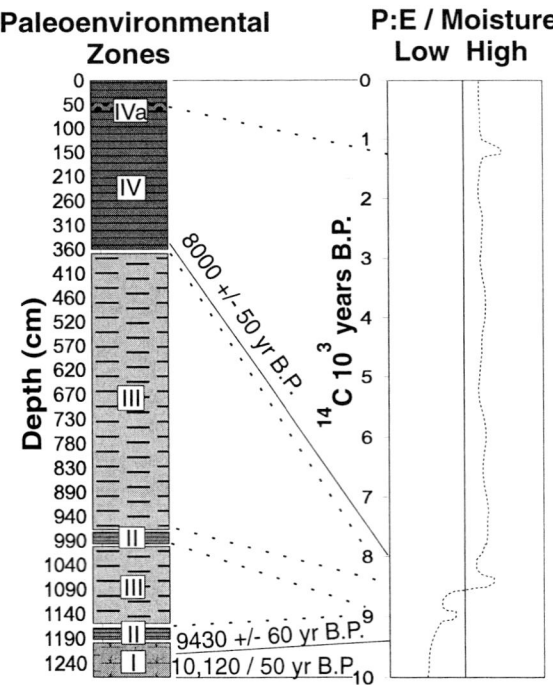

Figure 5.1 Paleoclimatic interpretation. P:E, inferred precipitation:evaporation ratio. Dashed lines are estimated ages.

69

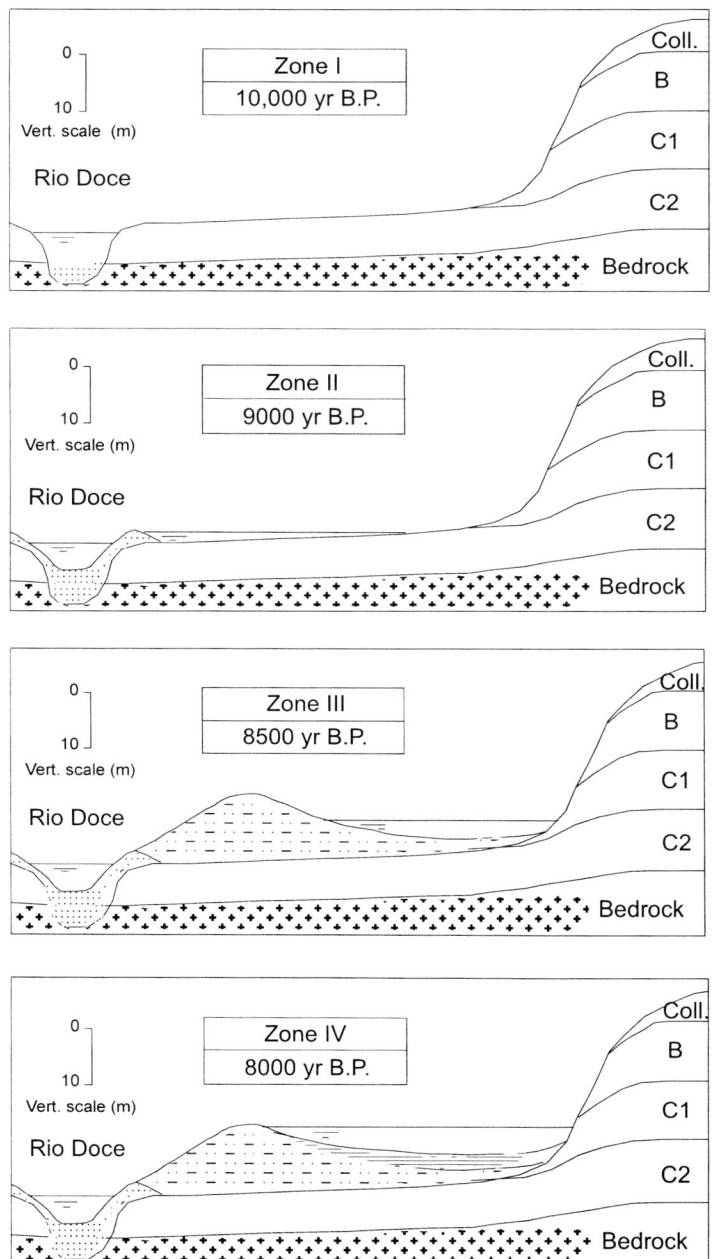

Figure 5.2 Schematic representation of the climate-driven agradation process that gave rise to the Lake Silvana basin (modified after Pflug, 1969).

Records of Pb, Cu, Cr and Hg concentrations show quite different behavior through the changing climatic and hydrological conditions from the beginning of zone IV. Hg and Cu concentrations, and to a certain extent Cr, remained at similar levels in zone IV relative to the underlying unit, while Pb concentrations became quite lower. As gibbsite contents also decreased, a raising lake level and the expansion of forest appear to have inhibited the erosion of all horizons of the weathering mantle (Figs. 4.3–4.5). Therefore, it is likely that Hg, Cu and partially Cr became more mobile and have been adsorbed on or co-precipitated with the authigenic component, as indicated by the positive correlation observed between Hg/Al ratios and Mn concentrations in zone IV (Fig. 4.8A). The hydrous Mn(III,IV) oxides are important mediators in the oxidation of oxidizable trace elements (Stumm and Morgan, 1996). It is well known that the redox cycling of Fe and Mn has relevant effects on the adsorption of trace elements under different redox conditions (Balistrieri et al., 1994; Song and Müller, 1999). Investigations on the behavior of trace metals in anoxic sediments of the Neckar River, Germany, revealed that particulate Cr and Co are simultaneously released from the dissolution of Fe and Mn oxides in the anoxic zone. The reprecipitation of Fe and Mn oxides in the overlying oxic zone can remove Cr and Co diffused from deeper sediments by adsorption or coprecipitation (Song and Müller, 1999). Another mechanism that could explain the relative enrichment of Cr in the authigenic component of zone IV has been indicated by a speciation study on dissolved trace elements in Hall Lake, USA (Balistrieri et al., 1994). It has been observed that the reduced form of Cr tends to form strong organic complexes in anoxic environments.

5.2 Siderite-inferred recurrence of climatic events

As the Rio Doce lakes present no thermal or chemical stratification during winter due to a complete circulation of the water mass, it can be realized that the history of siderite precipitation registered in sediments may provide important evidence of the climate-driven paleolimnological evolution in these lakes, as well as of water depth over time. Another study on the limnology of the Rio Doce lakes indicates an effect of the rainfall waters in the stabilization process of the water column, giving support for this hypothesis (Tundisi, 1997). The author attributes a very rapid increase in stratification and stability immediately after the beginning of the rainy season to the inflow of rainfall waters, which are enriched in nutrients and cooler than the epilimnion, so that they would reach the metalimnion and thus enhance the stratification process.

The association of siderite-rich layers with coarser sediments was interpreted as a result of episodes of increasing seasonality with abnormally strong summer rainfall, since seasonal variations in the water chemistry of several lakes of the Rio Doce basin show that chemical stratification reaches a maximum during the rainy summer (Tundisi et al., 1997). Such conditions are normally accompanied by concentrated input of terrestrial organic matter and Fe-rich clastics and enhanced chemical stratification in the lake, providing favorable conditions for siderite precipitation (Williamson et al., 1998; Xiouzhu et al., 1995; Moore et al., 1992; Postma, 1982). Rajan et al. (1996), based on a thermodynamic model for precipitation of siderite in the Black Sea, concluded that the precipitation of siderite from iron-rich waters is constrained by several thermodynamic variables including low pe, high CO_2 concentration and availability of organic matter.

Three distinct geochemical environments have been demonstrated to be conductive to siderite precipitation during early diagenesis. In slowly deposited pelagic sediments with relatively low concentrations of organic matter, vertically expanded depth zones of iron reduction constitute an ideal environment for siderite precipitation (Curtis and Coleman, 1986). The zone of methanogenesis constitutes a further highly favoring environment for siderite formation. Moreover, siderite can also coprecipitate with iron sulfides, but only if sufficient reactive iron is available and the rate of iron reduction balances or exceeds that of sulfate reduction (Curtis and Coleman, 1986; Walker, 1984). The latter environmental description seems to correspond to the conditions that have led to siderite precipitation in zone IV, which is accentuated in the sections 280-260, 180-160 and 80-40 cm.

The vertical distribution of the siderite-rich layers, arranged at intervals of ~1 m, points to the cyclic recurrence of sharp climate fluctuations following intervals of 2000-2500 years, assuming that sedimentation rates did not undergo significant variation within zone IV (Figs. 4.2, 4.5). Although a better chronological control is required to test this hypothesis, it is noticeable the possible correspondence in time with cyclic climatic events recorded from Greenland ice cores (O'Brien et al., 1995). A regular recurrence of peaks in the concentration of sea salt and terrestrial dust has been detected, pointing to cold and dry conditions during the periods 0 to 600, 2400 to 3100, 5000 to 6100 and 7800 to 8800 yr B.P. The latter is the same event that is believed to be associated with the increase in rainfall magnitude responsible for the deposition of zone III. The authors suggest that cooler climates recurred at quasi-2600-year intervals during the Holocene in the Northern Hemisphere. The interpretation of the causes of this pattern are still very speculative: a north polar vortex expansion or enhanced meridional circulation. Considering the possibility that the events of abnormally strong summer rains indicated for Lake Silvana have recurred contemporaneously to those in Greenland, it could provide important additional data for interpreting the nature of these events.

Authough an opposite interpretation has been proposed for the association of siderite with silty-sandy sediments in Brazilian lakes, that is a lower lake level leading to the deposition of coarser clastics during episodes of aridity (Absy et al., 1989; Soubiès et al., 1989), it is unlikely that such a hydrological change could result in increasing lake stratification and more reducing conditions in bottom waters. Moreover, the palynological record from Lake Silvana shows no evidence of forest retreat in the siderite-rich layers (Fig. 4.5). Finally, recent studies on lake sediments in New Zealand and Canada have demonstrated the association of storm events and increasing moisture, registered in historical climate records, with the deposition of coarse clastics eroded from the catchment (Eden and Page, 1998; Campbell, 1998).

6 Conclusions

The present study indicates the suitability of using geochemical and mineralogical records from lake sediments as paleoenvironmental proxies, at least for tropical regions, where the chemical weathering yields contrasting patterns among soil horizons with regard to their mineralogical and chemical composition. Chemical and mineral stratigraphy allows to a certain extent the reconstruction of the sedimentation history in Lake Silvana, whose main events are in close agreement with the climate-driven changes in vegetation registered in the palynological record. As concentrations of Hg, Pb, Cr, Ti and gibbsite show a strong variation within pedological profiles, their relative imprints in lake sediments have indicated changes in erosion activity on the catchment of Lake Silvana since the early Holocene. The recognition of pedological imprints in lake sediments has been favored by indicated oxic and sub-oxic conditions in the lake for phases preceding the last lake transgression, namely during the deposition of the zones I and III. Such conditions have led to prevailing deposition of detrital sediments which are composed of varying contents of kaolinite, gibbsite, oxyhydroxides, muscovite and quartz.

It is noteworthy that the observed accumulation of Hg at the top of weathering profiles, which seems to be a widespread characteristic of Brazilian soils, is probably not caused by a weathering-driven relative enrichment. The ionic potential of Hg and other observations from tropical regions suggest that Hg is likely to be leached rather than relatively enriched with increasing weathering degrees. It is suggested that the source of Hg in surficial lateritic soils is related to a long-term deposition of atmospheric Hg derived from natural emissions.

The onset of the last lake transgression, which coincides with the pollen-inferred expansion of tropical forest shortly before 8000 yr B.P., yielded increasing concentrations of organic matter on the lake bottom and consequently conditions of anoxia. This is reflected by the increasing relevance of authigenic Fe, Mn and siderite in zone IV. Relatively constant concentrations of Hg and Cr, considering the lower input of detrital sediments in zone IV, are probably a result of their mobilization under moister climates and through the formation of a denser vegetation cover.

With regard to the constraints that yielded the flooding of the lake towards the present levels, there are strong indications of a morphodynamic process caused by the rapid sedimentation of slope-wash sediments that culminated around 8500 yr B.P. with the damming of the drainage basin. The same process may be assumed for the formation of the whole lake system of Rio Doce. Intense slope erosion has been interpreted as a result of an abrupt climatic shift from dry to moister conditions accompanied by strong summer rains. This interpretation is supported by two pollen-indicated climatic transitions from dry to moister conditions that culminated 8500 years ago, whose age resembles that of expansion of forest registered in southern and southeastern Brazil and in southeastern Africa.

An early to mid-Holocene climatic event observed in records from ice cores, pointing to a notable increase in aridity in the Northern Hemisphere, suggests that the observations from Lake Silvana could be related to changes in the global climate. This hypothesis is also supported by the indicated cyclic recurrence of siderite-rich layers, arranged at intervals of ~ 1 m in zone IV. Strengthened conditions that favor siderite precipitation are believed to be related with increasing seasonality and abnormally strong summer rains. These episodes possibly recurred at intervals of 2000–2500 years, and if so, they correspond in time with cyclic climatic events of cold and dry climates recorded from Greenland ice cores. Thus, it is suggested that the very limited thermodynamic stability field of siderite may allow its use as a potential paleoenvironmental proxy for tropical lakes.

The present results suggest that sediment cores from the Rio Doce lakes may help a great deal in investigating the recent history of paleoenvironmental changes and their climatic implications in southeastern Brazil. Future works in this region are strongly recommended, where special attention to a high resolution of the sedimentary record should be paid, as well as to the dating of additional sections within the organic-rich sequence (zone IV).

7 References

Absy, M.L., Van der Hammen, T., Soubiès, F., Suguio, K., Martin, L., Fournier, M. and Turcq, B. (1989). Data on the history of vegetation and climate in Carajás, Eastern Amazonia. *In*: Proceedings of the International Symposium on Global Changes in South America during the Quaternary. INQUA, São Paulo, pp. 129-131.

Alley, R.B., Mayewski, P.A., Sowers, T., Stuiver, M., Taylor, K.C. and Clark, P.U. (1997). Holocene climatic instability: A prominent, widespread event 8200 yr ago. *Geology*, 25: 483-486.

Balistrieri, L.S., Murray, J.W. and Paul, B. (1994). The geochemical cycling of trace elements in a biogenic meromitic lake. *Geochimica et Cosmochimica Acta*, 58: 3993-4008.

Behling, H. (1993). Untersuchungen zur spätpleistozänen und holozänen Vegetations- und Klimageschichte der tropischen Küstenwälder und der Araukarienwälder in Santa Catarina (Südbrasilien). Dissertationes Botanicae Band 206, J. Cramer, Berlin-Stuttgart, 149 pp.

Behling, H. (1995). A high resolution Holocene pollen record from Lago do Pires, SE Brazil: Vegetation, climate and fire history. *Journal of Paleolimnology*, 14: 253-268.

Berner, R.A. (1981). A new geochemical classification of sedimentary environments. *Journal of Sedimentary Petrology*, 51: 359-365.

Bigarella, J.J. and Andrade-Lima, D. (1982). Paleoenvironmental changes in Brazil. *In*: G.T. Prance (ed.), Biological Diversification in the Tropics. Columbia University Press, New York, 714 pp.

Blunier, T., Chappellaz, J., Schwander, J., Stauffer, B. and Raynaud, D. (1995). Variations in atmospheric methane concentration during the Holocene epoch. *Nature*, 374: 46-49.

Braucher, R., Bourlès, D.L., Colin, F., Brown, E.T. and Boulangé, B. (1998). Brazilian laterite dynamics using in situ-produced ^{10}Be. *Earth and Planetary Science Letters*, 163: 197-205.

Campbell, C. (1998). Late Holocene lake sedimentology and climate change in southern Alberta, Canada. *Quaternary Research*, 49: 96-101.

Chesworth, W. (1992). Weathering systems. *In*: Martini, I.P. and W. Chesworth (eds), Weathering, Soils and Paleosols. Elsevier, Amsterdan. Developments in Earth Surface Processes, vol. 2, pp. 19-40.

COHMAP Members (1988). Climatic changes of the last 18,000 years: observations and model simulations. *Science*, 241: 1043-1052.

Costa, J.E. and Schuster, R.L. (1988). The formation and failure of natural dams. *Geological Society of America Bulletin*, 100: 1054-1068.

Coutinho, L.M. (1982). Ecological effects of fire in Brazilian cerrado. *In*: Huntley, B.J. and B.H. Walker (eds), Ecology of Tropical Savannas. Springer-Verlag, Berlin, Heidelberg. Ecological Studies, vol. 84, pp. 82-105.

Curi, N. and Franzmeier, D.P. (1984). Toposequence of oxisols from the Central Plateau of Brazil. *Soil Science Society of America Journal*, 48: 341-346.

Curtis, C.D. and Coleman, M.L. (1986). Controls on the precipitation of early diagenetic calcite, dolomite and siderite concretions in complex depositional sequences. *In*: Gautier, D.L. (ed.), Roles of organic Matter in Sediment Diagenesis. Society of Economic Paleontologists and Mineralogists, Tulsa. pp. 25-33.

DNPM (1984). Geologia do Brasil. Departamento Nacional da Produção Mineral, Brasília. 501 pp.

Drever, J.I. (1982). The Geochemistry of Natural Waters. Prentice Hall, New Jersey. 387 pp.

Eden, D.N. and Page, M.J. (1998). Palaeoclimatic implications of a storm erosion record from late Holocene lake sediments, North Island, New Zealand. *Palaeogeography,Palaeoclimatology, Palaeoecology*, 139: 37-58.

Engstrom, D.R. and Wright Jr., H.E. (1984). Chemical stratigraphy of lake sediments as a record of environmental change. *In*: Haworth, E.Y. and Lund, J.W.G. (eds). Lake Sediments and Environmental History. University of Minnesota Press, pp. 11-67.

Faegri, K. and Iversen, J. (1989). Textbook of Pollen Analysis (4th ed.). John Wiley & Sons, Chichester. 328 pp.

Fletcher, W.K. (1980). Handbook of Exploration Geochemistry: analytical methods in geochemical prospecting. John Wiley & Sons, New York. 255 pp.

Grimm, E.C. (1987). A Fortran 77 program for stratigraphically constrained cluster analysis by the method of the incremental sum of squares. *Pergamon Journals*, 13: 13-35.

Gunkel, G. and Sztraka, A. (1986). Die Bedeutung der Eisen- und Mangan-Remobilisierung für die hypolimnische Anreicherung von Schwermetallen. *Archiv für Hydrobiologie*, 106: 91-117.

IBGE (1993). Mapa de Vegetação do Brasil. Fundação Instituto Brasileiro de Geografia e Estatística, Rio de Janeiro.

IBGE (In Press). Projeto Radambrasil, Levantamento de Recursos Naturais - Folha Belo Horizonte. Fundação Instituto Brasileiro de Geografia e Estatística, Rio de Janeiro.

Irion, G. (1984). Sedimentation and sediments of Amazonian rivers and evolution of the Amazonian landscape since Pleistocene times. *In*: Sioli, H. (ed.), The Amazon: Limnology and Landscape Ecology of a Mighty Tropical River and its Basin. Dr. W. Junk Publishers, Dordrecht, pp. 537-579.

Johnson, T.C., Scholz, C.A., Talbot, M.R., Kelts, K., Ricketts, R.D., Ngobi, G., Beuning, K, Ssemmanda, I. and Mcgill, J.W. (1996). Late Pleistocene desiccation of Lake Victoria and rapid evolution of cichlid fishes. *Science*, 273: 1091-1093.

Jonasson, I. and Boyle, R.W. (1979). The biogeochemistry of mercury. *In*: Jaworski, J.F. (ed), Effects of Mercury in the Canadian Environment. National Research Council Canada, Ontario, pp. 28-49.

Jordão, C.P., Pereira, J.C., Brune, W., Pereira, J.L. and Braathen, P.C. (1996). Heavy metal dispersion from industrial wastes in the Vale do Aço, Minas Gerais, Brazil. *Environmental Technology*, 17: 489-500.

Kopp, D. (1986). Mineralogische, geochemische und geophysikalische Untersuchungen von Verwitterungsprofilen in Minas Gerais, Brasilien. Ph. D. Dissertation, Universität Freiburg, 213 pp.

Kronberg, B.I., Fyfe, W.S., Leonardos Jr., O.H. and Santos, A.M. (1979). The chemistry of some Brazilian soils: element mobility during intense weathering. *Chemical Geology*, 24: 211-229.

Lacerda, L.D. (1997). Global mercury emissions from gold and silver mining. *Water, Air and Soil Pollution*, 97: 209-221.

Ledru, M.P., Behling, H., Fournier, M., Martin, L. and Servant, M. (1994). Localisation de la forêt d'Araucaria du Brésil au cours de l'Holocène. Implications paléoclimatiques. *Comptes Rendus de'l Academie des Sciences Paris*, 317: 517-521.

Ledru, M.P., Bertaux, J., Sifeddine, A. and Suguio, K. (1998). Absence of Last Glacial Maximum records in lowland tropical forests. *Quaternary Research*, 49: 233-237.

Lucas, Y., Luizão, A., Chauvel, A., Rouiller, J. and Nahon, D. (1993). The relation between biological activity of the rain forest and mineral composition of soils. *Science*, 260: 521-523.

McFarlane, M.J. (1983). Laterites. *In* Goudie, A.S. and Pye, K. (eds), Chemical Sediments and Geomorphology. Academic Press, London, pp. 7-58.

Malm, O. (1998). Gold mining as a source of mercury exposure in the Brazilian Amazon. *Environmental Research*, 77: 73-78.

Martin, L., Fournier, M., Mourguiart, P., Sifeddine, A., Turcq, B., Absy, M.L. and Flexor, J.M. (1993). Southern Oscillation signal in South American palaeoclimatic data of the last 7000 years. *Quaternary Research*, 39: 338-346.

Matschullat, J., Gaber, U., Raphael, S. and Kober, B. (1998). Rekonstruktion der Versauerungsgeschichte eines Sees – sedimentologische, diatomologische und geochemische Untersuchungen an Sedimenten des Oderteiches im Harz. *Neues Jahrbuch für Geologie und Paläontologie Abhandlungen*, 208: 39-54.

Meis, M.R.M. (1977). As unidades morfoestratigráficas neoquaternárias do médio vale do rio Doce. *Anais da Academia Brasileira de Ciências,* 49: 443-459.

Meis, M.R.M. and Monteiro, A.M.F. (1979). Upper quaternary „rampas": Doce river valley, Southeastern Brazilian plateau. *Zeitschrift fürGeomorphologie,* 23: 132-151.

Meis, M.R.M. and Moura, J.R.S. (1984). Upper Quaternary sedimentation and Hillslope evolution: southeastern Brazilian plateau. *American Journal of Science*, 284: 241-254.

Meis, M.R.M. and Tundisi, J.G. (1997). Geomorphological and limnological processes as a basis for lake Typology. The middle Rio Doce Lake System. *In*: Tundisi, J.G. and Saijo, Y. (eds). Limnological studies on the Rio Doce Valley Lakes. Brazilian Academy of Sciences, University of São Paulo, pp. 25-48.

Merino, E., Nahon, D. and Wang, Y. (1993). Kinetics and mass transfer of pseudomorphic replacement: Application to replacement of parent minerals and kaolinite by Al, Fe, and Mn oxides during weathering. *American Journal of Science*, 293: 135-155.

Mitamura, O. and Hino, K. (1997). Distribution of biogeochemical constituents in lake waters. *In*: J.G. Tundisi, J.G. and Saijo, Y. (eds). Limnological studies on the Rio Doce Valley Lakes. Brazilian Academy of Sciences, University of São Paulo, pp. 97-108.

Modenesi, M.C. (1988). Quaternary mass movements in a tropical plateau: Campos do Jordão, São Paulo, Brazil. *Zeitschrift für Geomorphologie*, 32: 425-440.

Moore, S.E., Ferrel Jr, R.E. and Aharon, P. (1992). Diagenetic siderite and other ferroan carbonates in a modern subsiding marsh sequence. *Journal of Sedimentary Petrology*, 62: 357-366.

Müller, G. (1979). Schwermetalle in den Sedimenten des Rheins - Veränderungen seit 1971. *Umschau*. 79: 778-783.

Müller, G. and Gastner, M. (1971). The „Carbonate-Bombe", a simple device for the determination of carbonate contents in sediments, soils and other materials. *Neues Jahrbuch für Mineralogie Abhandlungen*, 10: 466-469.

Nahon, D.B. (1986). Evolution of iron crusts in tropical landscapes. *In*: Colman, S.M. and Dethier, D.P. (eds), Rates of Chemical Weathering of Rocks and Minerals. Academic Press, London, pp. 169-191.

Nahon, D.B. (1991). Introduction to the Petrology of Soils and Chemical Weathering. John Wiley & Sons, New York, 313 pp.

Nimer, E. (1989). Climatologia do Brasil. 2nd ed., Instituto Brasileiro de Geografia e Estatística, Rio de Janeiro, 422 pp.

Nishimura, M., Mitamura, O., Saijo, Y., Hino, K., Barbosa, F.A.R. and Tundisi, J.G. (1997). Geochemical information on biological sources of large amounts of sedimentary organic matter in four lakes. *In*: Tundisi, J.G. and Saijo, Y. (eds). Limnological studies on the Rio Doce Valley Lakes. Brazilian Academy of Sciences, University of São Paulo, pp. 169-188.

O'Brien, S.R., Mayewski, P.A., Meeker, L.D., Meese, D.A., Twickler, M.S. and Whitlow, S.I. (1995). Complexity of Holocene climate as reconstructed from a Greenland ice core. *Science*, 270: 1962-1964.

Parrish, J.T., Ziegler, A.M. and Scotese, R. (1982). Rainfall patterns and the distribution of coals and evaporites in the Mesozoic and Cenozcic. *Palaeogeography, Palaeoclimatology, Palaeoecology*, 40: 67-1С1.

Pflug, R. (1969). Das Überschüttungsrelief des Rio Doce, Brasilien. *Zeitschrift für Geomorphologie*, 13: 141-162.

Postma, D. (1982). Pyrite and siderite formation in brackish and freshwater swamp sediments. *American Journal of Science*, 282: 1151-1185.

Rajan, S., Mackenzie, F.T. and Glenn, C.R. (1996). A thermodynamic model for water column precipitation of siderite in the Plio-Pleistocene Black Sea. *American Journal of Science*, 296: 506-548.

Rodrigues-Filho, S. (1995). Metais Pesados nas Sub-Bacias Hidrográficas de Poconé e Alta Floresta (MT). CETEM/CNPq, Rio de Janeiro. Série Tecnologia Ambiental, vol. 10, 92 pp.

Rodrigues-Filho, S. and Maddock, J.E.L. (1997). Mercury pollution in two gold mining areas of the Brazilian Amazon. *Journal of Geochemical Exploration*, 58: 231-240.

Rose, A.W., Hawkes, H.E. and Webb, J.S. (1979). Geochemistry in Mineral Exploration. Academic Press, London, 658 pp.

Rosenbaum, J.G., Reynolds, R.L., Adam, D.P., Drexler, J., Sarna-Wojcicki, A.M. and Whitney, G.C. (1996). Record of middle Pleistocene climate change from Buck Lake, Cascade Range, southern Oregon - Evidence from sediment magnetism, trace-element geochemistry, and pollen. *Geological Society of America Bulletin*, 108: 1328-1341.

Roubik, D.W. and Moreno, J.E. (1991). Pollen and Spores of Barro Colorado Island. Missouri Botanical Garten, vol. 36, 270 pp.

Roulet, M., Lucotte, M., Rheault, I., Tran, S., Farella, N., Canuel, R., Mergler, D., and Amorin, M. (1996). Mercury in amazonian soils: accumulation and release. *In*: Proceedings of the Fourth International Symposium On The Geochemistry Of The Earth's Surface. Yorkshire, pp. 453-457.

Saijo, Y., Mitamura, O. and Barbosa, F.A.R. (1991). Chemical studies on sediments in the Rio Doce Valley Lakes, Brazil. *Verh. Internat. Verein. Limnol.*, 24: 1192-1196.

Salgado-Labouriau, M.L. (1973). Contribuicão à Palinologia dos cerrados. Publicacão da Academia Brasileira de Ciências, Rio de Janeiro, 291 pp.

Salgado-Labouriau, M.L., Casseti, V., Ferraz-Vicentini, K.R., Martin, L., Soubiès, F. Suguio, K. and Turcq, B. (1997). Late Quaternary vegetational changes in cerrado and palm swamp from Central Brazil. *Palaeogeography, Palaeoclimatology, Palaeoecology*, 128: 215-226.

Schindler, D.W., Turner, M.A., Stainton, M.P. and Linsey, G.A. (1986). Natural sources of acid neutralizing capacity in low alkalinity lakes of the Precambrian Shield. *Science*, 232: 844-847.

Servant, M., Soubiès, F., Suguio, K., Turcq, B. and Fournier, M. (1989). Alluvial fans in southeastern Brazil as an evidence for early Holocene dry climate period. Proceedings of the International Symposium on Global Changes in South America during the Quaternary. INQUA, São Paulo, pp. 75-77.

Song, Y. and Müller, G. (1999). Sediment-Water Interactions in Anoxic Freshwater Sediments: Mobility of Heavy Metals and Nutrients. Springer-Verlag, Berlin, Heidelberg. Lecture Notes in Earth Sciences, vol. 81, 111 pp.

Soubiès, F., Suguio, K., Martin, L., Leprun, J.C., Servant, M., Turcq, B., Fournier, M., Delaume, M. and Sifeddine, A. (1989). The Quaternary lacustrine deposits of the Serra dos Carajás (State of Pará, Brazil). Proceedings of the International Symposium on Global Changes in South America during the Quaternary. INQUA, São Paulo, pp. 125-128.

Stager, J.C. and Mayewski, P.A. (1997). Abrupt early to mid-Holocene climatic transition registered at the equator and the poles. *Science*, 276: 1834-1836.

Street-Perrott, F.A. (1993). Ancient tropical methane. *Nature*, 366: 411-413

Stumm, W. and Morgan, J.J. (1996). Aquatic Chemistry: Chemical Equilibria and Rates in Natural Waters. 3rd ed., Wiley-Interscience, New York, 1022 pp.

Tardy, Y., Melfi, A.J. and Valeton, I. (1988). Climats et paléoclimats tropicaux périatlantiques. Rôle des facteurs climatiques et thermodynamiques: température et activité de l'eau, sur la répartition et la composition minéralogique des bauxites et des cuirasses ferrugineuses, au Brésil et en Afrique. *Comptes Rendus de'l Academie des Sciences Paris*, 306: 289-295.

Tardy, Y. (1992). Diversity and terminology of lateritic profiles. In: Martini, I.P. and W. Chesworth (eds). Weathering, Soils and Paleosols. Elsevier, Amsterdam, pp. 379-405.

Tardy, Y. and Roquin, C. (1992). Geochemistry and evolution of lateritic landscapes. In: I.P. Martini, I.P. and W. Chesworth (eds). Weathering, Soils and Paleosols. Elsevier, Amsterdam, pp. 407-444.

Tintelnot, M. (1995). Transport and deposition of fine-grained sediments on the Brazilian continental shelf as revealed by clay mineral distribution (Ph.D. thesis, University of Heidelberg), 270 pp.

Tundisi, J.G. (1997). A note on the effect of rainfall in the process of stratification and stability in the Rio Doce lakes. *In*: Tundisi, J.G., and Saijo, Y., eds., Limnological studies on the Rio Doce Valley Lakes, Brazil: São Paulo, Brazilian Academy of Sciences/University of São Paulo, pp. 79-81.

Tundisi, J.G., Matsumura-Tundisi, T., Fukuara, H., Mitamura, O., Guillén, S.M., Henry, R., Rocha, O., Calijuri, M.C., Ibañez, M.S.R., Espindola, E.L.G. and Govoni, S. (1997). Limnology of fifteen lakes. *In*: Tundisi, J.G., and Saijo, Y., eds., Limnological studies on the Rio Doce Valley Lakes, Brazil: São Paulo, Brazilian Academy of Sciences/University of São Paulo, pp. 409-439.

Turcq, B., Pressinotti, M.M.N. and Martin, L. (1997). Paleohydrology and paleoclimate of the past 33,000 years at the Tamanduá river, Central Brazil. *Quaternary Research*, 47: 284-294.

Valero-Garcés, B.L., Laird, K.R., Fritz, S.C., Kelts, K., Ito, E. and Grimm, E.C. (1997). Holocene climate in the northern Great Plains inferred from sediment stratigraphy, stable isotopes, carbonate geochemistry, diatoms and pollen at Moon Lake, North Dakota. *Quaternary Research*, 48: 359-369.

Valeton, I. (1972). Bauxites. Elsevier, Amsterdam. Developments in Soil Science 1, 226 pp.

Veiga, M.M., Meech, J.A. and Oñate, N. (1994). Mercury pollution from deforestation. *Nature*, 368: 816-817.

Walker, J.C.G. (1984). Suboxic diagenesis in banded iron formations. *Nature*, 309: 340-342.

Williamson, D., Jelinowska, A., Kissel, C., Tucholka, P., Gibert, E., Gasse, F., Massault, M., Taieb, M., Van Campo, E. and Wieckowski, K. (1998).

Mineral-magnetic proxies of erosion/oxidation cycles in tropical maar-lake sediments (Lake Tritrivakely, Madagascar): paleoenvironmental implications. *Earth and Planetary Science Letters*, 155: 205-219.

Xiouzhu, Z., Yunfei, W. and Huaiyan, L., 1996, Authigenic mineralogy, depositional environments and evolution of fault-bounded lakes of the Yunnan Plateau, south-western China. *Sedimentology*, 43: 367-380.

Ybert, J.P., Turcq, B. and Albuquerque, A.L. (1996). Données préliminaires sur l'évolution paléoécologique et paléoclimatique holocéne dans la région moyenne du Rio Doce (Minas Gerais, Brésil) d'après l'analyse palynologique de deux carottes du lac Dom Helvacio. In: C.R. du Congrès Dynamique À Long Terme Des Écosystèmes Forestiers Intertropicaux. ORSTOM, Bondy, pp. 295-297.

Yu, Z.C. and Eicher, U. (1998). Abrupt climate oscillations during the last deglaciation in Central North America. *Science*, 282: 2235-2238.

8 Appendices

Appendix A.1 Metal concentrations in the sediment core SB1 – zone IV – partially measured by both analytical methods AAS and XRF with average AAS/XRF ratios, and total carbon and sulfur

Depth	Mn (μg/g)		Fe (%)			Al (%)			Si XRF	Ti XRF	K XRF	C	S
(cm)	X	SD	AAS	SD	XRF	AAS	SD	XRF	(%)	(%)	(%)	(%)	(%)
0	713	23	11.2	0.6	14.5	8.1	0.9	10.8	15.2	0.58	0.27	9.9	0.02
10	546		9.8		14.4	8.1		11.5	14.8	0.60	0.27	9.5	0.07
20	516	14	9.9	0.8	14.7	7.3	0.6	11.9	14.7	0.61	0.28	9.0	0.06
30	734.5	29.5	12.8	0.3	15.6	7.0	0.3	11.1	16.4	0.62	0.39	7.6	0.04
40	1293		16.1		15.6	6.1		10.5	17.3	0.62	0.45	4.7	0.05
50	2269		21.2			5.2						4.3	0.18
60	1696	25	13.5	0.4		6.6	0.8					4.3	0.08
70	1505		12.0		13.5	7.3		12.9	18.1	0.64	0.49	3.2	0.02
80	1387		14.3		15.5	6.2		11.1	17.9	0.63	0.46	4.1	0.07
90	941		12.5		15.5	7.7		12.4	15.4	0.66	0.39	6.0	0.14
100	852		12.5		16.1	6.8		12.4	15.2	0.66	0.31	6.8	0.10
110	763		11.7		15.8	7.2		12.6	14.6	0.65	0.31	6.3	0.10
120	906		12.9		18.1	6.8		11.7	13.6	0.62	0.27	7.5	0.12
130	687	17	11.8	0.7	16.3	7.6	0.4	12.1	13.5	0.64	0.21	8.9	0.10
140	585.5	13.5	10.8	0.4	15.8	7.1	0.3	12.4	13.6	0.63	0.19	8.6	0.08
150	787		12.9		18.6	7.6		11.8	12.7	0.62	0.19	7.7	0.07
160	664		11.6		16.5	7.7		12.4	13.7	0.64	0.18	7.1	0.06
170	925		13.1		19.7	6.8		11.5	12.4	0.64	0.24	6.5	0.07
190												6.3	0.10
200	576.5	23.5	8.6	0.3	15.2	8.2	0.5	12.8	14.7	0.70	0.22	6.2	0.13
210												6.2	0.11
220	758		9.4		11.3	9.1		14.7	15.3	0.80	0.17	4.9	0.08
230												5.1	0.12
240	579.5	29.5	8.8	0.1	11.4	8.7	0.9	15.1	15.8	0.83	0.22	5.0	0.16
250												5.3	0.13
260	704		9.8		13.6	7.8		13.9	14.6	0.83	0.16	5.9	0.10
270												5.1	0.10
280	673		9.0		16.6	8.2		13.4	13.6	0.84	0.12	5.6	0.10
290												4.7	0.09
300	594		8.6		15.3	9.3		14.5	14.7	0.88	0.14	3.8	0.09
310												4.2	0.07
320	571	35	8.9	0.3	14.1	9.0	0.4	15.1	14.9	0.95	0.13	4.1	0.06
330												3.4	0.04
340	293		8.3		13.5	9.9		16.1	15.4	0.95	0.18	2.0	0.03
350												1.3	0.02
360	205.5	18.5	5.9	0.5	10.4	13.3	0.3	17.3	16.4	0.87	0.27	0.3	0.01
Average AAS/XRF			0.72			0.62							
(standard deviation)			(0.12)			(0.05)							

X, AAS = average concentration;
SD = standard deviation of AAS duplicate.

Appendix A.2 Metal concentrations in the sediment core SB1 − zone IV − partially measured by both analytical methods AAS and XRF with average AAS/XRF ratios

Depth (cm)	Hg (µg/g)		Pb (µg/g)			Zn (µg/g)			Ni (µg/g)			Cu (µg/g)			Cr (µg/g)		
	X	SD	AAS	SD	XRF	AAS	SD	XRF	AAS	SD	XRF	AAS	SD	XRF	AAS	SD	XRF
0	0.92	0.06	36	5	41	142*	45	81	109*	5	74	59	4	55	63.5	9.5	172
10	0.28		29			109*			57			50			46		
20	0.22	0.04	22.5	3.5	32	112*	14	84	49.5	6.5	76	49	7	60	48	8	177
30	0.19	0.03	20	5	26	92*	6	79	49.5	8.5	68	45.5	3.5	49	47.5	9.5	141
40	0.22		26			99*			43			38			43		
50	0.38		47			147*			42			49			31		
60	0.43	0.04	60	6		96.5*	6.5		45	3		46.5	6.5		40	6	
70	0.25		38		28	150*		78	46		66	47		48	39		145
80	0.16		22		24	80*		76	46		65	40		45	45		133
90	0.20		33		26	101*		83	50		74	43		52	49		150
100	0.16		26		26	74		76	47		73	44		54	47		155
110	0.15		20		27	72		78	48		73	47		53	41		150
120	0.17		19		24	69		77	46		69	45		54	41		147
130	0.16	0.03	20	3	26	82	4	82	54.5	7.5	79	49.5	5.5	61	41	8	174
140	0.16	0.02	19.5	4.5	26	90	2	85	51.5	9.5	77	49	7	63	44		174
150	0.15		21		24	90		81	50		71	44		55	42		159
160	0.15		22		26	79		78	50		74	46		61	39		173
170	0.19		21		24	76		74	48		74	42		56	37		154
190																	
200	0.17	0.02	20.5	3.5	27	72.5	5.5	85	45	3	79	43	4	55	48.5	8.5	170
210																	
220	0.17		19		28	66		76	41		81	41		63	45		203
230																	
240	0.20	0.03	22	5	31	60	3	69	43.5	4.5	72	39.5	7.5	53	51.5	7.5	186
250																	
260	0.23		17		27	58		71	39		71	36		54	50		193
270																	
280	0.17		16		26	49		63	37		68	36		52	51		180
290																	
300	0.23		20		27	61		68	51		68	40		54	62		186
310																	
320	0.21	0.04	19	4	26	51	8	61	42	6	71	42	4	55	65	6	209
330																	
340	0.21		22		30	51		59	40		67	37		54	65		191
350																	
360	0.23	0.02	45.5	4.5	48	72.5	2.5	60	39	6	65	37	6	44	57	7	169
Average AAS/XRF			0.83			0.94			0.68			0.80			0.29		
(standard deviation)			(0.18)			(0.11)			(0.18)			(0.10)			(0.04)		

X, AAS = average concentration; SD = standard deviation of AAS duplicate;
* sample likely contaminated during digestion.

Appendix A.3 Major-elements-calculated mineral content – except for siderite – organic matter and granulometric distribution in the sediment core SB1 – zone IV

Depth	Kaoli-nite	Gibb-site	Goe-thite	Quartz	Side-rite	Musc./illite	O.M.	Total	<20µm	20-63	>63µm	>20µm
(cm)	(%)	(%)	(%)	(%)	(%)	(%)	(%)	(%)	(%)	(%)	(%)	(%)
0	40	5	20	12	4	4	15.8	101	72	21	7	28
10	43	5	21	10	4	4	15.0	102	55	14	31	45
20	43	5	21	10	4	4	14.2	101	76	3	21	24
30	40	4	22	14	4	5	12.0	101	76	3	21	24
40	34	5	20	14	6	13	7.4	99	66	5	29	34
50	27	5	22	10	14	15	6.8	100	7	1	92	93
60	40	3	18	10	14	8	6.8	100	2	1	97	98
70	50	3	15	13	7	6	5.1	99	67	11	22	33
80	50	0	19	13	6	6	6.3	100	47	24	29	53
90	51	3	20	7	6	5	9.5	102	60	23	17	40
100	50	3	23	7	4	4	10.7	102	69	21	10	31
110	50	4	24	6	2	4	10.0	100				
120	44	4	25	6	5	5	11.9	101	75	18	7	25
130	48	4	24	5	3	2	14.1	100				
140	50	4	23	5	3	2	13.6	101	92	6	2	8
150	46	4	25	5	5	2	12.2	99				
160	47	4	20	5	7	5	11.1	99	85	12	4	15
170	43	5	23	5	10	5	10.3	101	69	9	22	31
190	47	5	28	5	3	2	10.0	100				
200	54	3	24	5	3	2	9.8	101	83	13	4	17
210	55	5	15	5	5	5	9.8	100				
220	57	5	12	5	7	5	7.7	99	75	21	3	25
230	63	6	15	5	3	1	7.9	100				
240	60	6	16	5	3	2	7.9	100	88	11	1	12
250	51	3	22	7	5	3	8.4	99				
260	60	3	17	2	5	3	9.3	99	87	12	2	13
270	43	3	24	5	8	7	8.1	98	74	18	8	26
280	58	3	24	2	3	2	8.8	100	84	14	2	16
290	59	3	22	2	5	2	7.4	100				
300	61	4	22	2	3	2	6.0	100	84	13	3	16
310	64	3	21	2	2	2	6.6	101	83	14	3	17
320	64	2	19	1	2	2	6.5	96				
330	68	2	19	1	5	2	5.8	103	77	22	1	23
340	60	6	17	3	5	5	3.2	99				
350	66	6	11	3	5	5	1.6	98	90	9	2	10
360	60	10	12	4	5	7	1.3	99				

Appendix A.4 Metal concentrations in the sediment core SB1 – zones I, II and III – partially measured by both analytical methods AAS and XRF with average AAS/XRF ratios, and total carbon and sulfur

Depth	Mn (µg/g)		Fe (%)			Al (%)			Si XRF	Ti XRF	K XRF	C	S
(cm)	X	SD	AAS	SD	XRF	AAS	SD	XRF	(%)	(%)	(%)	(%)	(%)
380	155		5.0		9.1	13.9		18.3	16.8	0.96	0.29	0.75	0.001
400	182	12	5.8	0.7	8.6	13.3	0.5	18.6	17.1	1.03	0.27	0.68	0.001
420	183		6.2		8.9	13.8		18.9	17.0	1.03	0.25	0.82	0.004
440	248	16	6.9	0.3	9.3	13.0	0.9	19.1	17.3	1.03	0.27	0.71	0.002
460	295.5	20.5	6.9	0.4	9.3	13.8	0.4	18.7	17.3	1.05	0.25	0.71	0.001
480	268		6.8		9.2	12.7		18.5	17.1	1.05	0.28	0.71	0.001
510	274		7.0		9.7	12.3		18.6	16.4	1.06	0.26	0.62	0.001
530	299.5	14.5	7.1	0.8		13.3	0.2					0.69	0.001
550	207		6.6			12.6						0.72	0.001
570	295		6.9		9.4	12.8		19.0	16.9	1.05	0.26	0.79	0.001
590	337	21	6.7	0.4		12.0	0.7					0.79	0.001
610	352		6.7			12.7						0.69	0.001
630	380		6.5			12.1						0.89	0.002
650	316.5	18.5	6.4	1.1	9.8	11.7	0.3	18.1	16.1	1.1	0.26	0.94	0.001
670	340		6.5			12.6						0.81	0.001
690	349	15	6.6	0.5	9.7	13.6	0.8	18.6	16.4	1.04	0.23	0.71	0.001
710	352		6.8			13.2						0.63	0.001
740	451		6.7			12.8						0.86	0.001
760	314		6.4		9.6	13.5		18.8	16.7	1.02	0.22	0.65	0.002
780	375.5	22.5	6.7	0.3		11.2	0.2					0.60	0.001
800	327		6.4			12.5						0.59	0.002
820	303		6.3			12.1						0.65	0.004
840	332		6.6			11.9						0.69	0.005
870	282		6.3		9.5	10.6		19.7	17.8	1.08	0.25	0.70	0.005
890	318		6.4			11.3						0.70	0.004
910	425		6.7			11.9						0.65	0.002
930	294.5	19.5	6.7	0.7		12.1	0.4					0.64	0.004
950	307		6.5			12.3						0.65	0.006
970	311	18	7.0	0.8	14	12.0	0.3	16.9	15.4	1.03	0.11	1.05	0.005
990	484		11.2			11.3						1.54	0.002
1010	175		5.9			12.7						0.49	0.001
1030	161		6.5			12.6						0.38	0.005
1050	191		6.2			10.8						0.41	0.002
1070	198		5.9			12.0						0.46	0.005
1090	194	9	5.8	0.4	7.3	11.9	0.9	18.3	16.1	1.06	0.15	0.42	0.001
1110	144		5.2			11.8						0.41	0.001
1130	186		6.2			11.2						0.40	0.001
1150	219.5	12.5	5.6	0.6		11.2	0.3					0.42	0.001
1170	421		7.2			11.9						0.79	0.001
1190	449		7.3			11.2						0.93	0.002
1210	178		8.6		9.8	9.2		17.6	17.7	1.50	0.13	0.41	0.002
1230	153		6.7			9.2						0.64	0.002
1250	225	16	10.4	0.7		9.3	0.5					0.77	0.004
1270	118		7.7			8.8						0.74	0.005
Average AAS/XRF			0.70			0.68							
(standard deviation)			(0.09)			(0.07)							

X, AAS = average concentration;
SD = standard deviation of AAS duplicate.

Appendix A.5 Metal concentrations in the sediment core SB1 – zones I, II and III – partially measured by both analytical methods AAS and XRF with average AAS/XRF ratios

Depth	Hg (µg/g)		Pb (µg/g)			Zn (µg/g)			Ni (µg/g)			Cu (µg/g)			Cr (µg/g)		
(cm)	X	SD	AAS	SD	XRF	AAS	SD	XRF	AAS	SD	XRF	AAS	SD	XRF	AAS	SD	XRF
380	0.19		56		61	48		61	35		62	42		49	54		174
400	0.18	0.03	66.5	4.5	68	48	3	58	36.5	5.5	63	46	3	53	63	11	189
420	0.20		55		64	48		55	39		62	44		51	51		190
440	0.23	0.03	59	6		47	4		37	6		46.5	2.5		63.5	7.5	
460	0.19	0.04	55	8	70	48.5	2.5	56	37	4	66	42	4	56	55	4	198
480	0.20		56		67	46		60	35		65	43		57	52		183
510	0.16		57		67	50		58	33		63	42		52	50		195
530	0.19	0.02	59.5	6.5		50.5	3.5		37.5	7.5		43	6		54	6	
550	0.21		55			46			39			40			52		
570	0.24		58		70	49		55	41		63	45		51	65		186
590	0.22	0.04	55	4		46.5	4.5		35.5	5.5		40	3		52.5	5.5	
610	0.22		56			48			35			42			54		
630	0.51		55			60			33			45			52		
650	0.32	0.03	54.5	7.5	72	51	2	61	29	4	59	39.5	4.5	53	47	7	173
670	0.30		55			52			35			40			52		
690	0.26	0.05	60.5	4.5	73	53.5	3.5	58	38.5	4.5	62	44.5	3.5	53	70.5	8.5	194
710	0.31		59			47			33			40			54		
740	0.18		62			63			37			45			67		
760	0.18		57		72	48		57	36		59	39		52	57		178
780	0.20	0.04	54	8		48	6		34	5		42.5	5.5		68.5	4.5	
800	0.18		61			49			37			42			89		
820	0.20	0.03	55.5	6.5		46	3		35	6		40			75		
840	0.18		56			45			34			39			56		
870	0.21		52			40			28			35			47		
890	0.21		58			46			29			38			45		
910	0.21		63			55			32			42			53		
930	0.19	0.02	60.5	9.5		43.5	5.5		26	5		38	5		51	6	
950	0.20		65			51			35			36			37		
970	0.18	0.02	58	7	49	49	2	53	35.5	4.5	65	37.5	6.5	45	41	9	189
990	0.22	0.03	23	4		46			40			33			53		
1010	0.17		44			46			41			36			47		
1030	0.22		43			48			36			38			49		
1050	0.17		45			47			37			38			44		
1070	0.17		44			51			38			38			52		
1090	0.16	0.03	41	7		48.5	2.5		39.5	4.5		38.5	4.5		55	8	
1110	0.17		51			45			26			28			29		
1130	0.18		43			45			37			36			42		
1150	0.18	0.03	47	5		46	3		34.5	6.5		33	6		45	4	
1170	0.20		37			49			37			32			43		
1190	0.20		41			55			37			34			42		
1210	0.11		51		56	46		48	47		65	64		72	105		273
1230	0.10		30			35			30			38			36		
1250	0.08	0.02	47.5	7.5		43	2		36	4		52.5	3.5		97.5	6.5	
1270	0.09		32			33			29			52			94		
Average AAS/XRF				0.88			0.86			0.59			0.82			0.30	
(standard deviation)				(0.11)			(0.05)			(0.06)			(0.05)			(0.05)	

X, AAS = average concentration; SD = standard deviation of AAS duplicate.

Appendix A.6 Major-elements-calculated mineral content – except for siderite – organic matter and granulometric distribution in the sediment core SB1 – zones I, II and III

Depth	Kaoli-nite	Gibb-site	Goe-thite	Quartz	Side-rite	Musc./illite	O.M.	Total	<20µm	20-63	>63µm	>20µm
(cm)	(%)	(%)	(%)	(%)	(%)	(%)	(%)	(%)	(%)	(%)	(%)	(%)
370									69	26	4	31
380	60	13	10	5	5	7	1,2	101				
390									43	28	29	57
400	66	10	9	5	5	5	1,1	101				
410									39	38	23	61
420	63	11	9	6	5	5	1,3	100	37	40	23	63
440	63	14	9	6	5	5	1,1	103	42	43	15	58
460	62	13	9	6	5	5	1,1	101	47	36	17	53
480	58	14	10	7	5	5	1,1	100	39	52	9	61
510	57	15	11	6	5	5	1,0	100	43	45	12	57
530							1,1		45	44	11	55
550							1,1		50	42	8	50
570	56	16	9	7	5	7	1,1	101	42	46	12	58
590							1,2		51	47	2	49
610							1,1		56	41	3	44
630							1,4		45	45	10	55
650	58	15	11	5	5	5	1,5	101	48	45	7	52
670							1,3		46	45	9	54
690	59	15	11	5	5	5	1,1	101	36	39	25	64
710							1,0		50	41	9	50
740							1,4		32	40	28	68
760	63	12	11	4	5	5	1,0	101	49	40	11	51
780							0,9		40	37	23	60
800							0,9		35	42	23	65
820							1,0		47	47	6	53
840							1,1		46	44	10	54
870	65	12	9	3	5	5	1,1	100	52	39	9	48
890							1,1		53	43	4	47
910							1,0		45	43	12	55
930							1,0		51	40	9	49
950							1,0		50	45	5	50
970	55	10	14	7	10	5	0.9	102	46	51	3	54
990							2,4		75	24	1	25
1010							0,8		69	29	2	31
1030							0,6		62	37	1	38
1050							0,6		43	51	6	57
1070							0,7		58	32	11	42
1090	63	12	10	5	5	5	0,7	101	54	32	14	46
1110							0,6		60	38	8	40
1130							0,6		47	47	6	53
1150							0,7		59	40	1	41
1170							1,2		90	10	0	10
1190	60	10	10	5	8	5	1,5	100	95	5	0	5
1210	80	1	15	1	1	1	0,6	99	57	40	3	43
1230							1,0		64	28	8	36
1250							1,2		51	39	10	49
1270							1,2		59	29	12	41

Appendix A.7 Metal concentrations in the weathering profiles EG1 and EG2 partially measured by both analytical methods AAS and XRF with average AAS/XRF ratios

Profile/ Horizon	Depth (m)	Hg (µg/g) X	SD	Pb (µg/g) AAS	SD	XRF	Zn (µg/g) AAS	SD	XRF	Ni (µg/g) AAS	SD	XRF
EG2/Coll.	0.1	0.25	0.03	3.5	1.5		19.5	0.5		15.5	3.5	
EG2/Coll.	4.0	0.28		5		11	22		23	17		23
EG2/B	7.6	0.02	0.01	12.5	2.5	16	8	1	24	2	1	10
EG2/B	10.1	0.07	0.02	9	1		12.5	1.5		8	2	
EG2/B	10.4	0.01		15		19	12		19	4		10
EG2/B	10.9	0.01	0.00	11	2		11.5	2.5		2.5	0.5	
EG2/B	11.6	0.01		7			13			4		
EG2/B	12.0	0.04		13		14	11		26	3		10
EG2/B	13.0	0.01		13		16	8		15	4		9
EG2/B	14.0	0.01		20		16	11		43	12		42
EG2/B	14.2	0.01		8			11			10		
EG2/C1	14.4	0.03	0.01	102	5	90	19	2	23	5	1	15
EG2/C1	24.0	0.03		87		61	14		17	12		20
EG2/C2	25.0	0.02	0.01	42	6	37	137	4	125	100	6	113
EG2/C2	29.0	0.03		30		31	52		52	44		58
EG1/Coll.	0.5	0.22		10			28			30		
EG1/Coll.	1.0	0.18	0.02	9.5	1.5		23.5	1.5		15.5	3.5	
EG1/Coll.	2.0	0.15		8			26			14		
EG1/Coll.	3.0	0.13		9			52			20		
Average AAS/XRF					0.97			0.69			0.49	
(standard deviation)					(0.26)			(0.28)			(0.22)	

X, AAS = average concentration; SD = standard deviation of AAS duplicate.

Profile/ Horizon	Depth (m)	Mn (µg/g) X	SD	Cu (µg/g) AAS	SD	XRF	Cr (µg/g) AAS	SD	XRF	Fe (%) AAS	SD	XRF	Al (%) AAS	SD	XRF
EG2/Coll.	0.1	22	4	12.5	0.5		42	5		3.8	0.3		11.2	0.5	
EG2/Coll.	4.0	27		10		4	48		156	3.8		5.1	11.1		13.6
EG2/B	7.6	12	2	6	1	1	1.5	0.5	15	2.0	0.1	2.2	2.7	0.3	13.4
EG2/B	10.1	23	2	7.5	0.5		16	1		3.1	0.2		4.5	0.2	
EG2/B	10.4	31		11		6	2		12	2.7		3.7	3.9		12.1
EG2/B	10.9	14.5	2.5	3.5	0.5		5.5	2.5		2.2	0.3		2.9	0.5	
EG2/B	11.6	25		4			5			0.8			2.6		
EG2/B	12.0	30		6		1	8		27	2.9		2.7	4.6		12.7
EG2/B	13.0	6		7		1	3		18	1.9		2.5	4.3		13.1
EG2/B	14.0	5		9		1	6		14	0.4		0.6	5.8		15.6
EG2/B	14.2	1		5			5			1.3			3.5		
EG2/C1	14.4	35	4	14	2	16	409	12	800	11.6	0.6		3.1	1.2	
EG2/C1	24.0	49		16		12	178		337	6.9		7.7	3.7		13.8
EG2/C2	25.0	67.5	3.5	35.5	2.5	39	53	4	100	21.8	1.1	23.9	9.7	0.6	11.0
EG2/C2	29.0	62		44		56	18		60	15.4		19.1	6.7		11.9
EG1/Coll.	0.5	143		45			75			8.9			12.3		
EG1/Coll.	1.0	113	5	32	2		73.5	5.5		8.1	0.4		11.9	0.2	
EG1/Coll.	2.0	133		35			75			9.8			11.1		
EG1/Coll.	3.0	285		51			70			10.3			11.4		
Average AAS/XRF					3.62			0.33			0.83			0.46	
(standard deviation)					(2.90)			(0.15)			(0.12)			(0.23)	

X, AAS = average concentration; SD = standard deviation of AAS duplicate.

Appendix A.8 Major-elements-calculated mineral content in the weathering profile EG2

Profile/ Horizon	Depth	Si XRF	Ti XRF	K XRF	Kaoli-nite	Gibb-site	Goe-thite	Hema-tite	Quartz	Musco-vite	Total
	(m)	(%)	(%)	(%)	(%)	(%)	(%)	(%)	(%)	(%)	(%)
EG2/Coll.	0.1				33	19	8	0	37	0	97
EG2/Coll.	4.0	25,1	0,54	0,01	35	19	8	0	37	0	99
EG2/B	7.6	29,8	0,22	0,03	64	0	0	2	34	0	100
EG2/B	10.1				60	0	0	3	37	0	100
EG2/B	10.4	29,8	0,33	0,05	58	0	0	5	37	0	100
EG2/B	10.9				61	0	0	5	34	0	100
EG2/B	11.6				63	0	0	3	34	0	100
EG2/B	12.0	28,5	0,31	0,01	61	0	0	5	33	0	99
EG2/B	13.0	28	0,25	0,01	63	0	0	4	31	0	98
EG2/B	14.0	26,9	0,05	0,29	72	0	0	1	22	5	100
EG2/B	14.2				68	0	0	2	28	0	98
EG2/C1	14.4				55	0	6	14	21	0	96
EG2/C1	24.0	25,8	0,73	0,00	65	0	3	7	24	0	99
EG2/C2	25.0	13,2	1,79	0,01	52	0	24	14	4	0	94
EG2/C2	29.0	16,8	1,52	0,00	57	0	24	7	8	0	96

95

Appendix A.9 Summary diagram of pollen taxa - Sediment core SB1

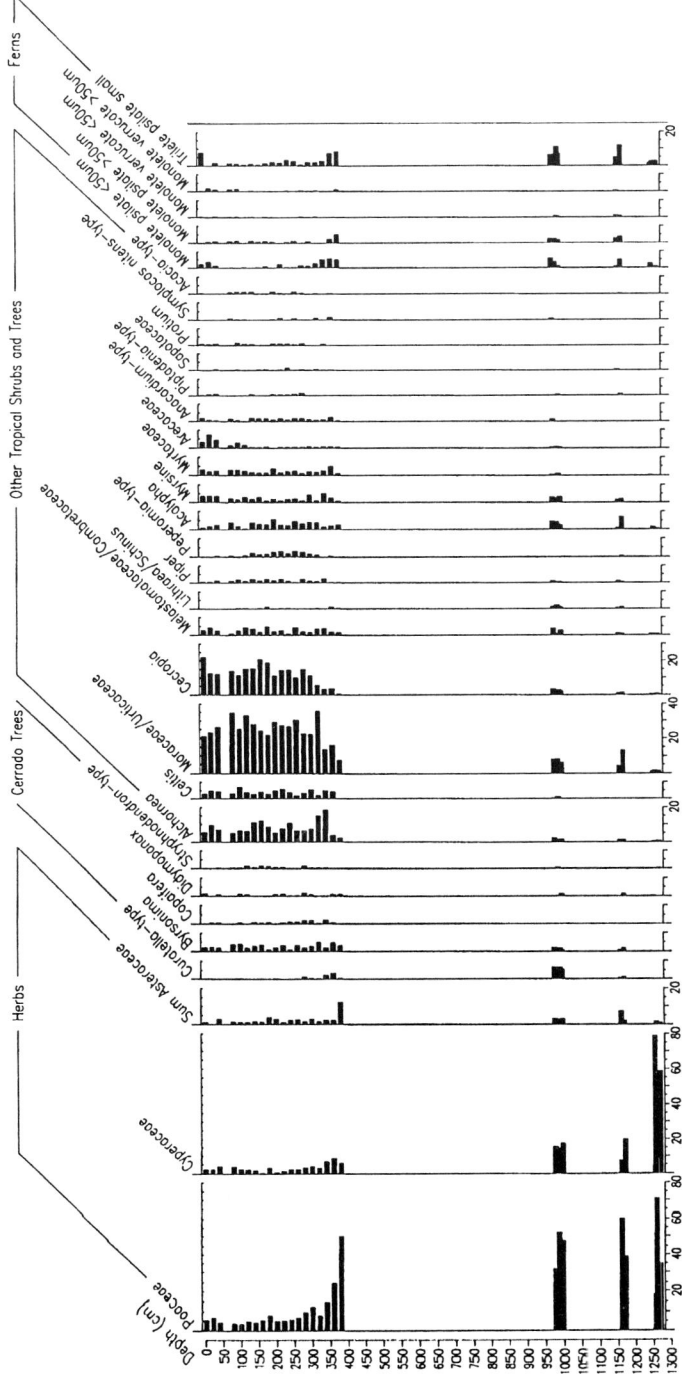

96

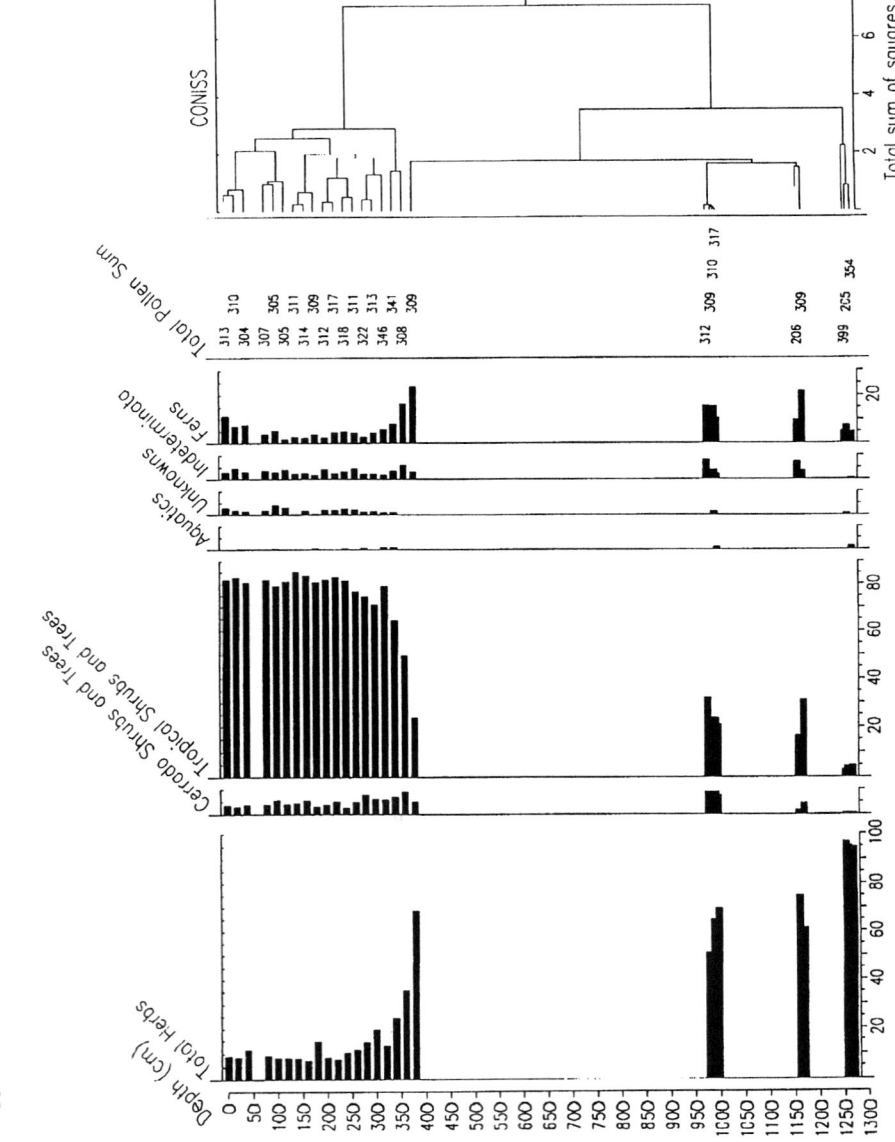

Appendix A.10 Summary pollen diagram with ecological groups and a cluster analysis dendrogram

Lecture Notes in Earth Sciences

For information about Vols. 1–19
please contact your bookseller or Springer-Verlag

Printing: Druckhaus Beltz, Hemsbach
Binding: Buchbinderei Schäffer, Grünstadt